D1294181

FROM ARISTOTLE TO DARWIN AND BACK AGAIN

From Aristotle to Darwin and Back Again

A Journey in Final Causality, Species, and Evolution

ETIENNE GILSON

Translated by John Lyon

UNIVERSITY OF NOTRE DAME PRESS
NOTRE DAME

Originally Published in 1971 as
D'Aristote a Darwin et retour
by Librairie Philosophique J. Vrin

Library of Congress Cataloging in Publication Data

Gilson, Étienne, 1884-1978.
From Aristotle to Darwin and back again.

Translation of: D'Aristote à Darwin et retour.
Includes bibliographical references and index.
1. Biology – Philosophy. 2. Evolution – Philosophy. I. Title.
QH331.G4713 1984 574'.01 83-40114
ISBN 0-268-00967-8

Manufactured in the United States of America

We always fall back upon, we revolve in a certain circle around, a small number of solutions which have been mutually related and mutually antagonistic from the beginning. It is customary to be astonished that the human mind is so unlimited in its combinations and reach; I modestly confess that I am astonished that it is so limited.

Sante Beuve, *Portraits litteraires*
(Pleiade, II, p. 466.)

"Why – damn it – it's *medieval,*" I exclaimed; for I still had all the chronological snobbery of my period and used the names of earlier periods as terms of abuse.

C.S. Lewis, *Surprised by Joy*
[London: Geoffrey Bless, 1955
(chapter XIII, p. 195).]

Contents

Foreword

THAT TRANSLATION IS THE ART of betrayal, or that the translator is the traitor, is a thought that can never be far from the mind of one who attempts to "bring across" meaning from one language to another. It is a particularly frightening thought for one who is not *ex professo* a philosopher or a linguist when he is engaged in translating a work such as the present one, by an author as professionally accomplished in philosophy as M. Gilson was. I can only plead in extenuation of my "betrayal" that the task needed to be done, plus the traditional exigencies of the amateur – the lover, in this case, of wisdom.

I have had an open flirtation with the history and philosophy of science for some time now, and have been particularly engaged by the complex (and at times seemingly disingenuous) rhetoric of Darwin's arguments in *The Origin of Species* and *The Descent of Man.* I have also been greatly taken by the concept of final causality, although it be perhaps more in its Platonic form than in its Aristotelian. (I am thinking primarily of the argument about the primacy of mind in *Laws,* Book X, and the moving definition of mind in the *Phaedo* as "that which orders all things for the best," along with the subsequent suggestion that meaning is contingent upon our "saving the appearances" – to borrow another Platonic phrase – or showing how that which is phenomenally observable can be explained in morally and aesthetically acceptable terms.) Explanation through citing the necessary preconditions of an event, or explanation by "proximate" or physical causality alone, seems hollow. A world without ultimate or metaphysical causes is a silly world. These factors drew me to M. Gilson's work.

The initial problem the translator faces with this work is how to render the key word *finalité* in English. After much thought, and after having gone through the entire work several times, I decided to allow the context in which the word would have to be embedded in English to determine how the word would best be translated. Only when the context prevents the word from appearing too blatant a neologism is "finality" or "finalism" used in English. Otherwise, the word is generally translated as "teleology" or "final causality," as its context seems to demand. I have usually allowed "finalism" and "finalist" to stand in English rather than try to replace them with some necessarily awkward phrase. These two words ordinarily occur in close proximity to either "teleology" or "final causality," as in the chapter on Bergsonism, for instance, and their meaning is never in doubt. (The Mitchell translation of Bergson's *Creative Evolution* uses "finalism" also, I should note.)

My debts in connection with this translation are innumerable, yet some must be mentioned. Father Stanley L. Jaki suggested this work to me, read the entire translation, and has been a constant support to me throughout the extended period that became necessary in seeing this book to press. I owe him an extraordinary debt of gratitude. My colleague and one of my dearest friends, Professor Bernard Doering, saved me from many a disastrous mistranslation. My former colleague, now Brother Thomas Frerking, O.S.B., of St. Louis Priory, also read the typescript and gave me much valuable advice. I owe a debt of gratitude to Dean Robert Burns of the College of Arts and Letters at the University of Notre Dame for a faculty development grant which allowed me to begin this translation—so many summers ago now. Dr. Walter Nicgorski, chairman of the Program of Liberal Studies at the University of Notre Dame, has been a constant shepherd of my often wayward talents, and I owe him particular recognition and thanks. Mr. James Langford of the University of Notre Dame Press, and Mr. John Stoll Sanders of Henry Regnery–Gateway, have each been cooperative in this venture. Mr. John Ehmann of the Notre Dame Press worked tirelessly at the thankless task of editing my typescript.

Penultimately, I wish to thank my undergraduate assistants John Scanlon and Patrick Farris for their cheerful and intelligent help in the initial and transitional stages of preparing the typescript. Finally, I have only the highest praise for the devotion, skill, and prayerful sensitivity of my research assistant, Laurie Lee Tychsen, who remained with me throughout the six most difficult months of my life which happened to occur while this translation was being put in its final form, though she most certainly was severely tempted to do otherwise.

I have "called my shots" as I have seen them in this translation, and no doubt I have seen some of them too faintly, or perhaps even misconceived my target completely in the dark. Where I have been right, I often must share the praise with those who have helped me. Where I have erred, I stand alone.

John Lyon
St. Mary's College
Winona, Minnesota
August 1983

Introduction

EARLY IN HIS CAREER Etienne Henri Gilson realized that he had a passion for writing. He put down the pen only a year or so before he died in 1978 at the age of ninety-four. After his seventy-fifth birthday in 1959, when the list of his publications comprised well over six hundred entries, Gilson added to that list another hundred or so titles. It is into that latter group that the work here translated belongs.

Written by him in his mid-eighties, this work may easily take first place among books written by octogenarians. The youthfulness of its author's intellectual vigor exudes from each page. It is not infrequent that a great mind preserves its penetrating power to a very old age. Quite different is the case with respect to that openness to the latest which is a hallmark of this book. Moreover, the book is a witness to Gilson's being knowledgeable about the latest in a field, biology, which had never been his profession. All his life Gilson had been a philosopher who searched for truth with an eye on the history of philosophy, and in particular with an eye on the history of medieval philosophy.

His teaching career never strayed from that field once it really began. After World War I, in which he was cited for heroism, he was appointed as professor of the history of philosophy at the University of Strasbourg. By 1921 he was at the Sorbonne as professor of medieval philosophy, and ten years later he took the same chair in the Collège de France, the highest distinction which France can offer to any of her children active in teaching. Gilson's election to the Académie Française took place in 1947. Parallel with recognition in France there came such foreign acknowledgments of Gilson's preeminence as his serving repeatedly in the late 1920s as visiting professor at Harvard University, his giving the Gifford Lectures at the University of Aberdeen in 1930 and 1931, and the William James Lectures at Harvard in 1938. By then he had been a chief force behind the rapid growth of the Pontifical Institute of

Medieval Studies in Toronto, which he joined as full professor in 1951. It was from that institute that Gilson's influence reached its zenith. He was a chief attraction to a large number of talented young scholars who, following their studies with him, became the heralds of their master's vision in a great number of American universities.

The vision included far more than the particulars of the scholarship of a specialist. Gilson was never a mere historian. Had he been one, he would not have experienced more than a keen surprise when as a doctoral student of Levy-Bruhl at the Sorbonne he discovered a curious facet in Descartes' writings. While Descartes heavily relied on the phraseology of the Scholastics, he gave a new twist to many of their pivotal terms. Gilson was enough of a philosopher to see not only the crucial importance of Descartes' tactics for the rest of Western intellectual development but also its lesson for man's perennial quest for truth or intellectual certainty. It was in the wake of making that discovery that Gilson, who went through a Catholic lycée around 1900 without ever hearing a word about Thomas or Thomism, perceived what he firmly believed to be the forever valid philosophical truth of a truly perennial philosophy.

The two works, *The Spirit of Medievel Philosophy* (Gifford Lectures) and *The Unity of Philosophical Experience* (William James Lectures), in which Gilson shared with a broad readership the philosophical truth as he held it, are too well known to call for a comment here. Less known for the general reader is the position that Gilson staked out for himself (with Maritain being a close ally) within Thomism. His position had two chief characteristics. One was his emphasis on Thomism as a Christian philosophy. Whatever the full rationality of Thomistic philosophy, it was in a crucial sense, so Gilson kept arguing, the fruit of Christian consciousness steeped in the tenets of the Creed. If philosophy was, like any other reality, a reality existing in history, Thomism had to be comprehensible only as a reality rooted in its own history. The latter began with a history-making reflection of the medieval Christian consciousness on the non-Christian philosophical and scientific corpus, embodied mainly in Aristotle's works. Unlike many Thomists among his contemporaries, too keen to appear as thinkers free of any "prejudice," even if it were awareness of the historical origin of their philosophy, Gilson always wanted to be known as a Christian philosopher. In the present work too Gilson makes, whenever necessary, a clear reference to the intellectual rights of Christian revelation and theology.

The other chief characteristic of Gilson's position as a Thomist was also related to the historical reality of any trend or thing. Not that Gilson had ever advocated historicism, let alone its almost inevitable sequel, relativism. He proudly called himself a dogmatist, in the sense that he knew that he had perceived truths which, if truly such, had to be valid for any phase of history – intellectual, social, and biological. Truth perceived as unchangeable dogma was one thing, its effective articulation another. In the latter respect Gilson had again found himself separated from a large number of Thomists. The latter, in spite of their good intentions, all too often failed to look beyond the intellectual horizons of Thomas' century. Time and again they overlooked the lesson of the youthfulness with which Thomas seized on the latest (the thirteenth-century Latin discovery of Aristotle) in order to assure freshness to the perennial. Four decades before Gilson wrote the book here translated, he had warned his fellow Thomists that the freshness of perennial truths can only be secured by looking for new illustrations of it, and he pointed at modern science as their principal source.

Once in retirement and free of the duties of teaching (he never spared himself of being entirely at the disposal of his students) Gilson had the leisure to show what that second characteristic should look like when properly implemented. His book on modern linguistics (*Linguistique et philosophie: Essai sur les constantes philosophiques du langage*, 1969) is still to be translated into English, although it has already seen a reprinting (1981). That Gilson chose that subject shows something of his awareness of the principal areas where the great battle for the human mind is being waged in modern times. As a devout worshiper of the Word Made Flesh he had no illusion about the threat posed to Christian convictions when search for truth is replaced by a study of words and phrases separated from their bearing on reality as the embodiment of intelligibility.

For Gilson those convictions had to be far more than a vaguely emotional or esthetic matter. They had rather to be a fully reasoned service of the Christian Creator, who is the Logos Himself. Since that Logos is the source of full meaning, which is inconceivable without a concomitant purpose, Gilson, not surprisingly, saw the other chief area of that battle in the debates created by modern biological and evolutionary science about the purposiveness of life. This is why the central contention of the present book is the soundness of natural teleology, that is, a philosophical discourse about purposiveness in all living beings, not only in man whose consciousness is an irrefragable witness to it. That teleology, Gilson argues here, can at best be ignored by biologists but never systematically evaded without incurring the burden of blatant inconsistencies.

Gilson did not live to hear the president of the Royal Society, Sir Andrew Huxley, with impeccable Darwinist credentials warn his mostly Darwinist colleagues against taking lightly the question of the origin of life and, what is worse, against shoving under the carpet the even greater problem of the origin of consciousness and reasoning. (Quite a few of them are in fact animated by that rank materialism which sets the tone of Darwin's early notebooks, published in an unabridged and unexpurgated form only after Gilson wrote the present book.) Most reprehensibly, they acted in a philosophically cavalier manner even in relation to the pivotal notions of their subject matter, such as species, evolution versus e-volution, natural selection, survival of the fittest, organization of parts into a whole, and the like. At times, to recall only one of Gilson's inimitable phrases, they engaged in a "massacre of the universals" which, of course, robustly survive their repeated wholesale executions.

Chapters II, III, and IV of *From Aristotle to Darwin and Back Again* are an unraveling of that "scientific" manhandling of logic, a story stretching from Buffon through Lamarck beyond Darwin and Spencer into, in fact, the years postdating this book. Recent proposals about replacing the word teleology with teleonomy (as if the solution of philosophical problems were a mere matter of semantics) and about the formulation of a "biological metaphysics" would no doubt have been eagerly seized upon by Gilson. He would have found in those proposals further illustrations of the fact that those who frown most on metaphysics are most likely caught in its pseudoversions. Gilson's storytelling is all the more convincing as he once again proved himself to be that consummate matador who, in full familiarity with his target's most vulnerable spots, is able to make comments that are weighty for all their brevity and light touch.

But Gilson's chief aim was never to vanquish his opponents, gracefully as he could do this. Truth being his principal concern, he had to be mainly positive. Chapters I, V, VI, and VII, through a presentation of the teleology of Aristotle, of Bergson, and of some modern men of science, show him articulating the constants of biophilosophy which doggedly call for the biophilosopher's attention. These constants are the unfailing presence of facts, such as the baffling organization of all living beings from nonheterogeneous parts and their coherence in the whole across the flow of time, which invariably raise the specter of purposiveness. Without yielding ever so slightly to the temptation of facile anthropomorphism, which time and again discredited teleology, Gilson drives home his main point: only the doctrine of substantial form as understood by Aristotle, and not as misconstrued by Descartes and many moderns after him, can provide a basis for a teleology which is satisfactory for the philosopher and indispensable for the kind of biologist who is sensitive to the deeper meaning of his findings and interpretations.

During the last eight years of his life Gilson saw no comment of any consequence made by biologists on this book which is a more penetrating analysis of a culturally crucial topic than are the rows of books written on the history of purposiveness with an eye on evolution. While the neglect of this book by biologists may be excused on the ground that they hardly ever really care for matters philosophical, a similar neglect by historians and philosophers of science is another matter. A not too oblique remark of Gilson's in the first chapter of the volume here translated on a studious disregard for evidence may very well be applicable to some below-the-surface motivations at work in those professions. That in Catholic circles the book provoked no echo, let alone sustained studies, may be symptomatic of a vision in which a craft steadily losing altitude is seen by its enthused passengers as soaring toward ever new heights. Had Gilson's book been carefully and widely studied, those heights would have ceased being contemplated through lenses ground in workshops in which the tools of science, philosophy, and theology (and of poetry in prose) are blissfully interchanged to provide the pinkish hue of a facile cosmic optimism.

Something akin to the above suggestion was Gilson's considered judgment about many new developments carrying the label of Thomism, let alone developments within it whose spokesmen expressed disdain for that label. Unlike many other elderly thinkers, Gilson did not grow bitter or discouraged on seeing the rapid decrease of the influence of his lifelong work and message. He kept his courage and good cheer because he was that well-informed thinker who saw the constants in the welter of bewildering change. That biophilosophy has through its long and checkered history been driven back with breathtaking regularity, and often in spite of the grim resolve of some of its celebrated spokesmen, to its constants, so many pointers to natural teleology, was for him one aspect of a broader intellectual landscape where the battle is waged between two searches for certainty. One is the search in terms of a scientism which loses itself at regular intervals in the morass of skepticism after having staked everything on the certainties allowed by the scientific method. Gilson does not discuss scientism as such in this book, though he presents it from the very start as his obvious target. For only when scientism is seen for what it is, together with the cult accorded to it in modern Western culture up to this very day, can another search for certainty prevail. This is the search outlined in Gilson's methodical realism. There, because the methodical realist never tries to begin the intellectual march with the second or third step (the hallowed program of scientism and of philosophies patterned on science), the step to reality remains always the first step to take. Only when reality is approached this way can it serve as the source of freshness and wonder, which are, in Gilson's eyes, the unfailing marks of true philosophy.

Gilson never entertained any illusion about the possibility of transmitting the sense of wonder by philosophical discourse, however articulate and conclusive. Such a discourse is this book. In view of the Gallic finesse, of which hardly a paragraph written by Gilson is void, its rendering into English must have demanded no small skill and scrupulous care on the part of the translator, to whom the English-speaking world must indeed feel indebted. That world today wields an enormous influence on the intellectual orientation of mankind and does so for better or for worse. The infatuation of the English-speaking world with an evolutionary philosophy disdainful of purpose is hardly something to cheer about or something that would be cured by a book however excellent. But this book will be a source of intellectual strength for those in that world who refuse to be unmindful of purpose, individual and cosmic, and are ready to join the ongoing battle for its saving vision.

Stanley L. Jaki
Distinguished University Professor
Seton Hall University

Preface

THE NOTION OF FINAL CAUSALITY has not been treated kindly. One of the principal reasons for the hostility toward it is its long association with the notions of a creator God and providence. Already in the *Memorabilia*, I, 4, 5–7, Xenophon attributed to Socrates the idea that the intelligence of man could only be the work of an intelligent demiurge such as the one which, in the *Timaeus*, Plato would soon charge with the task of constructing the world. From then on the proof of the existence of God through final causality was never to leave theology. Whether through hostility to the notion of God; or through a desire to protect scientific explanation against all theological contamination, even though it be from natural theology; or whether, finally, through an alliance of these two motives, the representatives of what can be called "scientism" today agree upon the proscription of the notion of final causality.

We have no intention of discussing scientism. That is the resolve not to admit, in any discipline, any solution to any problem which cannot be rigorously demonstrated by reason and is not verifiable by observation. The object of the present essay is not to make of final causality a scientific notion, which it is not, but to show that it is a philosophical inevitability and, consequently, a constant of biophilosophy, or philosophy of life. It is not, then, a question of theology. If there is teleology in nature, the theologian has the right to rely on this fact in order to draw from it the consequences which, in his eyes, proceed from it concerning the existence of God. But the existence of teleology in the universe is the object of a properly philosophical reflection, which has no other goal than to confirm or invalidate the reality of it. The present work will be concerned with nothing else: reason interpreting sensible experience – does it or does it not conclude to the existence of teleology in nature? It is not certain that every truth concerning nature is scientifically demonstrable: Scientific demonstration as well as reason may not have anything valid to say about what experience indemonstrably suggests. Thus understood, the existence of natural teleology appears to be one of these philosophical constants whose inexhaustible vitality in history can only be recorded.

The philosopher who deals with such a problem constantly feels troubled in conscience by reason of his scientific incompetence in a matter where science is directly concerned. It is consequently a great satisfaction for him to come across occasionally a biologist who is aware of the existence and the nature of the philosophical problem posed by the organization of living beings. We shall take the liberty, then, of citing the testimony of Lucien Cuénot, of the Academy of Sciences, on the precise issue which will be the object of our own book. "The more one penetrates deeply into determinisms, the more do the relations become complicated; and as this complexity leads to a univocal result which the least deviation can disturb, then there is born invincibly the idea of an ultimate ordering [of things]. I admit that it is incomprehensible, indemonstrable, that it is to explain the obscure by the more obscure, but it is necessary. It is more especially necessary to the degree that one becomes better acquainted with determinisms, because it is impossible to do without a guiding thread in the cloth of events. It is not foolhardy to believe that the eye is made for seeing."

By different routes the present work leads to the same conclusion. This conclusion, it will be said, is not then original? No, it is only true, and it can be useful to repeat that in a time when it is good philosophic form to claim the contrary. One reads in the *Cahier de notes* of Claude Bernard: "Science is revolutionary." I am profoundly convinced that philosophy is not.

CHAPTER I

Aristotelian Prologue

AMONG THE WORKS OF ARISTOTLE the one least familiar to philosophers is the *History of Animals.* Scientists consider it to be scientifically outmoded, and philosophers do not consider it as philosophical in the modern sense of the word. It is, however, unquestionably Aristotelian, which suggests that the manner in which Aristotle himself conceived science and philosophy is not exactly the same as ours. In fact, although Aristotle did not hold himself to be a scientist, in the sense in which the scientist is a specialist in some branch of the sciences of nature, but only to be a man reasonably informed about the science of his times, such is too exalted a view for the taste of today's philosophers. Even those among them who read Aristotle are only slightly interested in his philosophy of nature. Those of our contemporaries who know proportionately as much about zoology and biology as Aristotle did are not professors of philosophy. They rather teach introductory courses in zoology and ecology in some preliminary university course of instruction.[1] His scientific curiosity appears to have come to life once more in Albert the Great, who possessed in the highest degree this typical gift of the born biologist, which is the taste for personal observation; but Thomas Aquinas, as so many others, does not appear to have considered this sort of information to be necessary *ad pietatem*, and he neglected it. Today zoologists and philosophers no longer speak to one another. For the researcher and the professor qualified in one of these disciplines, anyone else is simply ignorant of the facts. What modern professor of philosophy has ever spoken to his students of the teeth of dogs, horses, men, and elephants? Aristotle did so. His philosophy included, along with many other parts, this part of the science of his time.

As early as the first chapter of the *History of Animals* Aristotle invokes one of the numerous notions which can be held to be constants of the philosophy of nature, and which, moreover, as with most notions of this sort, is at one and the same time scientific and philosophical: the notion of homogeneity.

The first phrase of this treatise says that of the parts of which animals are composed, some are simple, others composite. Those which are simple divide themselves into parts of a uniform nature. For example, flesh is made of pieces of flesh. Composite parts divide themselves into parts which are not uniform among themselves. Thus, for example, "the hand does not divide into hands nor the face into faces."[2] If one calls the first class of parts homogeneous and the second class heterogeneous, one will have at one's disposal a distinction the scientific and philosophical consequences of which are still today of interest for the problem of final causality.

Among the generalities with which Aristotle tarries at the beginning of the *Parts of Animals* and which is relevant to our own inquiry, it is appropriate to indicate still another: the ancient authors, Aristotle says, were interested first of all in the process of formation of each animal,[3] which today would be called ontogenesis; but it is perhaps as important to consider the same animals once formed, "for there is no small difference between these two points of view."[4] Aristotle does not appear to have thought of calling these two methods of approaching living beings "diachronic" and "synchronic," but it is indeed what he is thinking of. He himself preferred first to describe completely formed animals, and only subsequently to describe the process of their formation. We shall see the connection of this preference with his doctrine of teleology.

A third remark is that, of the two kinds of parts which we have distinguished in living beings, the homogeneous and the heterogeneous, the second necessitates that one should take into consideration a peculiar type of causality. Different kinds of causes are at work in nature: the material, the formal, the efficient [*le moteur*], and the final. All whose structure is homogeneous can be explained by the efficient cause, which Aristotle frequently calls "the point of origin of motion." Heterogeneous parts require in addition, for their explication, another kind of cause, that which we call today the "final cause," and which Aristotle calls simply "the end" (*telos*), the "in view of which" (*to ou eneka*), the "why" (*dia ti*). Never does he use an abstract expression such as "final cause," and "finality" he uses even less. He speaks of real objects or of elements of these objects which may be as real as they.[5] If there is in the real a principle of unity–substance, for example–it is necessary that the four kinds of causes be able to return, in one manner or other, to this principle; a cause of any kind whatsoever *is* such only through it.[6]

Why is there heterogeneity in the structure of certain beings? Because they are living beings. A living being is a being which is born, grows, develops, comes to maturity, and, finally, through a process in the reverse direction, declines and dies. The living being then recognizes itself in this thing that changes, and as all change is motion, the order of the living is the order of motion. More precisely, it is that order of all which has in itself the principle of its own change. In abstract terms one says that the living being is endowed with spontaneity, not only in its reactions, but a fortiori in its operations and its actions.

That the living being moves itself entails as a consequence that it is composed of heterogeneous parts. Indeed, to move oneself consists in having in oneself the cause of one's movement. The living being is at the same time cause and effect, but it cannot be the one and the other in the same way. Aristotle expressly contradicts the Platonist notion which makes of life a simple source of motion, as if one single and identical thing could be motive force and thing moved at the same time and in the same way. It suffices to see an animal move about to ascertain that the parts which move take their point of departure from the fixed and the immobile. All living operations, all the growth of plants or animals, involve and require the differentiation of certain parts capable of acting one on another. Heterogeneity of parts is required for the very possibility of that causality operating on itself which characterizes the growth of living beings.

For the same reason it is necessary that the heterogeneous parts of the living being make up a certain order. The notion of order is inseparable from that of causality, which is itself an order of dependence. That which is cause under a certain aspect can be effect under another. The ability of a living being to move itself, even though it be only to assimilate and grow, involves therefore the organization of the heterogeneous parts of which it is composed. This is why one says of living bodies that they are organisms or that living matter is organic [*organisée*]. The finalism of Aristotle is an attempt to give a reason for the very existence of this organization.

Aristotle is often reproached for his anthropomorphism, that is to say, for his habit of considering nature from man's point of view. If to do so is an error, the reproach is justified, but Aristotle's attitude in this regard had nothing naive in it. He was conscious of it, just as he was of the reasons there are for adopting it. At the moment he begins the study of the parts of animals, he declares straightforwardly: "to begin with, we must take into consideration the parts of man. For, just as each nation reckons by that monetary standard with which it is most familiar, so must we do in other matters. And, of course, man is the animal with which we are all of us the most familiar."[7]

At first sight there is something disconcerting in this naivete. It seems far too simple to evaluate the parts of other animals in terms of those of the human body, as one evaluates foreign currency in terms of francs or dollars. Upon reflection, however, there is something to be said in favor of this proposition, for in a certain sense it is true. It is not necessarily that man may be better known to us than the rest [of creation], but, to begin with, whatever object is considered, the knowledge that we have of it is human knowledge which expresses itself in some human language; and, next, the knowledge which man has of himself, imperfect as it may be, is by nature privileged. In knowing himself man knows nature in a unique way, because in this unique case the nature that he knows, he is. In and through the knowledge which man has of himself nature knows herself directly; she becomes conscious of herself in him, self-conscious one might say, and there is strictly nothing else that man can hope to know in this way. Even other men, with whom he can communicate by language or any other sort of signs, remain for him parts of the "external world." In fact, all the rest of the universe is and remains for him the external world. Since then there is no other knowledge for each of us other than our own knowledge, things known exist for us only in relation to ourselves, and among these things there is only one that we can apprehend directly in itself, and that is what we are and what each calls "I," "me."

Fortified by his principle, Aristotle proceeds in a methodical manner from man to nature in his exploration of reality. The problem of the "end" in nature is for him only one more occasion for applying this method, which he holds to be universally valid. In the present case, that of the relation of homogeneous parts to heterogeneous parts in living bodies, Aristotle will first remark, as a thing immediately obvious, that homogeneous parts cannot themselves be composed of heterogeneous parts; such a supposition would be absurd. Faces, let us say, and limbs are composed of flesh; flesh is not composed of faces or of limbs. From that flows an important consequence.

Insofar as it is a question of problems where the parts involved are all homogeneous, matter is the sole cause to take into consideration, for matter itself is homogeneous. At this level mechanical explanations by matter alone account for reality in a satisfying manner. Beings of heterogeneous structure, on the contrary, require a more complex mode of explication. The heterogeneity of their component parts necessitates that they necessarily have structure, and the question presents itself whether the existence of such structures is susceptible of the same kind of material explanation which works so remarkably well in the case of homogeneous beings.

Especially since Descartes reduced the order of bodies, living or not, to that of pure geometrical extension according to the three dimensions of space, the problem no longer arises. The natural tendency of scientific explanation is to take for granted that the same kind of explanation will succeed in both cases. Assuredly the task is not without difficulties, the principal of which is to give an account of the structure of living beings—which, as we said, is a new fact in relation to the inorganic—without making a new principle of explanation intervene. To explain heterogeneous parts by the same principles which explain homogeneous parts is to leave deliberately unexplained the heterogeneity of the heterogeneous.

Yet this was tried before the time of Aristotle. Certain of our contemporaries, better informed about their science than about its history, readily imagine that because they themselves exclude it from scientific explanation to the benefit of mechanism, teleology is an ancient view and mechanism a modern one. Their history begins with Descartes, who inaugurated mechanistic physiology, which is consequently modern, by eliminating the finalist biology of Aristotle, a biology consequently ancient and, according to them, definitively superceded. But the contrary is true. Even if it were proved, biological mechanism would be a return to a view of the living being older than teleology, a view that Aristotle himself, the finalist, held to be definitively superceded. This is not only true of mechanism as a principle of explanation in general; in fact, the very responses imagined by Aristotle's predecessors in order to resolve the particular problem of the origin of the heterogeneous parts of organized beings curiously foreshadow, at least in the spirit which inspires them, those which are proposed today. When he was asked to explain the formation of the vertebrae, Empedocles tranquilly responded that they were constrained in the too restricted space which the fetus occupied in its mother's womb. Vertebrae are pieces of bone which were initially continuous but are presently broken.[8] Most often, the responses given to questions of this kind by the predecessors of Aristotle were luck, chance, "accidental encounters," or even necessity, but never foresight, design, or "the end."

Aristotle was not without arguments against this mechanism, and, according to his custom, they were arguments of common sense. First, Empedocles neglected the fact that living beings are not the results of chance; they come from seed endowed with definite formative properties whose products are themselves determined. In addition, it should not be forgotten that in all species the parents come before their offspring and predetermine their future development. Parents are not abstract principles, but real beings. Man is engendered, not by "chance," or by an "accidental encounter," but by a man.[9]

Returning to this point, Aristotle generalizes the problem and adds: "The same statement holds good also for the operations of art. . . . The products of art, however, require the pre-existence of an efficient cause homogeneous with themselves, such as the statuary's (sic) art, which must necessarily precede the statute; for this cannot possibly be produced spontaneously. Art indeed consists in the conception of the result to be produced before its realization in the material."[10] To which he adds that even if beings were produced by chance or came about as the result of a sort of spontaneity, mechanism would remain incapable of giving a reason for their production. The spontaneity of nature, and even chance, can indeed bring about the restoration of health; but in cases of this kind, the true cause of the restoration of health is neither spontaneity nor chance but rather nature itself, which circumstances have allowed to exercise its functions, or which they have stimulated to do so.

As serious as the other arguments may have appeared to Aristotle's mind, it is the last which appears to him to be decisive. To the question "How does nature produce beings made up of heterogeneous parts?" he responds by another question: "How does man fabricate objects made up of such parts?" Art imitates nature; it must be then that nature proceeds in a manner analogous to that of art.

That which comes first in the operation of art is the presence in the mind of the artist of a certain image or notion of the object to be produced. From that point of departure the artist begins by choosing material adopted to the structure of the future work. These would be, for example, heterogeneous parts: canvas, colors, and so on necessary to produce the particular picture which the painter has in mind. This necessity is a hypothetical necessity, the cause of which is the idea of the future picture already present to the mind of the painter. *If* the picture to be painted is such-and-such, then the constituent elements must necessarily be such-and-such.[11]

This is only an example, and the domain of the fine arts is not the only place where it is true. Artisans proceed in the same fashion as artists: all fabrication presupposes the image, concept, or idea of the object to be fabricated. Moreover, the order of action calls attention to our problem as much as that of production does. Except when it is a question of habitual acts, all that we do ought to be first foreseen, calculated, conceived before being executed. More simply, there ought to be a "reason" for what we do. Without such a preliminary notion in our mind nothing happens. This notion is the cause, for it is that without which something else would not exist. Since its causality consists in being the term or objective of the operation, one says that it is its "end." And since it is the presence of that end in thought which brings into action all the operations required to attain it, including the choice of and organization of means, it is called the first and principal cause, the "cause of causes," for it is the reason which it is necessary to allege in order to explain what a thing is and that it is such as it is. Departing from that point, and drawing an inference once more from man to other parts of nature, Aristotle concludes that if one asks which cause is primary, the material or mechanical cause, or the final cause, the response must be, "Plainly, however, that cause is the first which we call the final one." Then he adds, "For this is the reason for the existence of the thing, and the reason for the existence of things forms the starting point, alike in the works of art and in the works of nature."[12]

In Aristotle's mind it was less a question of a process of reasoning than a matter of fact. We see teleology, for we see beings constituted according to a certain order and a certain plan, with the result that species exist whose charateristics are constant, as if the future of these beings had been predetermined in the seed from which they were born. However, as soon as one thinks about it, the notion of the end becomes obscure. One asks oneself how it could be that something which does not yet exist could direct and determine that which already is, though it be only to conduct its operations or direct its growth.

In the case of man, whether it is a question of the order of production or that of action, the problem allows of a solution. Because he is endowed with consciousness, man can conceive the not-yet-existent end in view of which, in order that it might be attained, certain antecedent conditions must be fulfilled. There is no experience more common to us than that: all our active life is made up of such an enchainment of means and ends. Our speculative life is furthermore of the same character: logical operations and scientific research obey the primacy of the ends. Such is not the case with natural teleology. If nature operates in view of ends, neither the philosopher of nature nor the scientist can say in what mind these ends are first conceived. All takes place for them as if no ends of this kind intervened in the production of natural beings and of their structures. They feel as if bound by a sort of intellectual obligation, not only not to allow the end to intervene in their explanations but even to deny its existence, which they do as if their assumption were self-evident. But sometimes they proceed with an aggressive violence which can hardly be explained if purely speculative interests are the cause of the denial.

Aristotle had a clear consciousness of the difficulty, but, unlike certain of our contemporaries, a fact remained a fact for him even when he realized that he was incapable of explaining it. Speaking of beings composed of heterogeneous parts and endowed with a structure, he thought that whatever was the explanation of the facts, the science of nature ought to take these facts into consideration. The terms in which he expressed his opinion on this point deserve consideration. Assuredly, what he called the elements of nature were quite simple things in comparison with the living cells to which, even regardless of their life, modern biology is tempted to reduce organisms. For him everything was composed of the possible combinations of four elements – earth, water, air and fire – as also the exchanges between their elementary qualities: the dry and the humid, the cold and the hot. That was too simple, but Aristotle could repeat his argument today, as far as its basic structure is concerned, without having to modify it:

> For to say what are the ultimate substances out of which an animal is formed, to state, for instance, that it is made of fire or earth, is no more sufficient than would be a similar account in

> the case of a couch or the like. For we should not be content with saying that the couch was made of bronze or wood or whatever it might be, but should try to describe its design or mode of composition in preference to the material; or, if we did deal with the material, it would at any rate be with the concretion of material and form. For a couch is such and such a form embodied in this or that matter, or such and such a matter with this or that form; so that its shape and structure must be included in our description. For the formal nature is of greater importance than the material nature.[13]

We have intentionally allowed Aristotle to follow his own way, which, starting from the consideration of a living animal, leads him naturally to that of a fabricated object, such as a couch, as if the production of a natural object and that of an object made by the hand of man were for all purposes identifiable. Aristotle thinks that they are so, except, however, in this, that the teleology of nature is much more perfect than that of art. The artist gropes about, corrects himself many times, and often fails in his work, whereas nature, although she also may fail through the fault of matter, in general attains her end without hesitation.

Let us elaborate this point a bit, for the true nature of Aristotle's anthropocentrism, and consequently that of his finalism, are here involved.[14] We say that Aristotle imagines nature as a sort of artist who deliberates and makes a choice among appropriate means toward the end which he proposes to himself. And such a scheme is true in a sense, as we come to view it. But it is still more true to say that in the last analysis Aristotle conceives the artist as a particular case of nature. This is why, in his natural philosophy, art imitates nature, rather than nature imitating art. The contrary is imagined because, every man being more or less an artist, an artisan, and a technician, we know, more or less confusedly, yet with certitude, the manner in which art operates. On the contrary, insofar as we are natural beings, we are the products of innumerable biological activities of which we know practically nothing, or very little. The manner in which nature operates escapes us. Her finality is spontaneous, not learned. She normally operates with remarkable sureness, as is visible in the operation of instinct by which final causality external to the being adapts it to its surroundings, or in internal operations which show the mutual adaptation of the parts of animals in view of their end. In nature the end, the *telos*, works as every artist would wish to be able to work: in fact, as the greatest among them do work, or even the others in moments of grace when, suddenly masters of their media, they work with the rapidity and infallible sureness of nature. Such is Mozart, composing a quartet in his head while writing down its predecessor. Such is Delacroix, painting in twenty minutes the headpiece [*chapeau*] and mantle of Jacob on the wall of Saint-Sulpice. A technician, an artist who worked with the sureness of a spider weaving its web or a bird making its nest would be a more perfect artist than any of those that anyone has ever seen. Such is not the case. The most powerful and the most productive artists only summon from afar the ever-ready forces of nature which fashion the tree, and, through the tree, the fruit. This is why Aristotle says that there is more design (*to ou eneka*), more good (*to eu*), and more beauty (*to kalon*) in the works of nature than in those of art.

The analogy with art, then, assists us to recognize the presence in nature of a cause analogous to that which is intelligence in the operations of man, but we do not know what this cause is. The notion of a teleology without consciousness [*connaissance*] and immanent in nature remains mysterious to us. Aristotle does not think that this should be a reason to deny its existence. Mysterious or not, the fact is there. It is not incomprehensible because of its complexity, which we can only hope science will one day clarify, but because of its very nature, which does not allow it to be expressed in a formula.

What is it then that the modern biologist wishes to say by declaring that it is *scientific* to exclude final causality from the explanation of organized living beings? In connection with an analogous case, imposture has recently been spoken of.[15] The word would not be properly applied to our case. Imposture is the act of an imposter, and an imposter is he who imposes upon others through false appearances or mendacious talk. Sometimes genuine fraud is practiced in science, but it is alien to the enterprise, is extremely rare, and in the present case the deceit could profit no one. It is difficult to imagine a scientist denying the validity of the notion of the final cause for the sole pleasure of leading others into error. Apart from the fact that the idea does not make sense, it contradicts what one knows of the true scientist and his unconditional respect for the truth, which is the mainspring of all his scientific activity, and consequently of his personal moral life.

For a scientist to be able to say something like that, it would have to be that it appeared to him to be a scientific statement. He is not trying to deceive others; he simply deceives himself. The pure mechanist in biology is a man whose entire activity has as its end the discovery of the "how" of the vital operations in plants and animals. Looking for nothing else, he sees nothing else, and since he cannot integrate other things in his research, he denies their existence. This is why he sincerely denies the existence, however evident, of final causality.

That costs him something. A sort of intellectual asceticism alone allows the scientist [*scientiste*] to deny the evidence which each moment of his life as man, and even of his activity as scientist [*savant*], does not cease to confirm. One can, moreover, doubt that the most resolute adversaries of finalism can succeed in eliminating it from their mind. The occasions for coming back to this point are numerous. Let us at least note at present that it is difficult to speak of the *function* of an organ or of a tissue without dangerously brushing against the idea of a natural teleology. To say of a machine or a mechanical accessory that they function, or that they "run," implies the notion that they function as they ought to function and as it had been foreseen that they would function. If a machine or any apparatus whatsoever does not fulfill the function for which it has been built, it is simply thrown away. The same is true in biology, and particularly in medicine. An ailing heart is a heart which does not function as it ought. Ever since Claude Bernard the biological identity of the normal and the pathological has been rightly affirmed. A generalized cancer destroys an organism in a completely natural fashion, one conformable to the laws which regulate the development of viruses. The organism lives and dies conformably to the same laws, but we do not adjudge normal everything that happens in nature, and the normal organ is that whose structure is in such a state that it assures the proper functioning of the organ. The biologist perhaps does not put the question to himself as to the "why" of the living body which he studies, but he can nevertheless not avoid ascertaining that, in fact, if the structure of an organism alters in a serious fashion, it will cease to exist. One reinstates the organism, then, in the condition in which it ought to be *in order* that it might exist, for an animal is normally a living being. Even at this superficial and completely empirical level it is difficult to avoid the notion of final causality. But the issue will be pursued at greater length in due course.

Further along on this subject the position of Aristotle is again instructive. As early as the fifth century before Christ, Democritus had identified the essence of animals with the form of their bodies, their exterior aspect and their color. This was not absurd, and up to a certain point was even true. To be a horse or an oak is, in the first place, to be for us a living being which, to judge of them by their external appearance, we recognize as belonging to these species. That which we recognize them to be ought, therefore, to be that which they are. If such were truly the opinion of Democritus (Aristotle does not appear to be entirely sure of it, moreover), he conceived of the nature of living beings as defined by their configuration, without any recourse to the notion of final causality. But is this possible? If animals were only their visible forms, there would be no difference between the living and the dead. But these are entirely different from one another. A dead man resembles a living one as much as can be, yet he is no more a man but a corpse. "And yet a dead body has exactly the same configuration as a living one; but for all that is not a man."[16] After which, giving way once more to the inclination of his considered anthropomorphism, Aristotle observes that the

> physiologists, when they give an account of the development and causes of the animal form, speak very much like such a craftsman [a woodcarver]. What, however, I would ask, are the forces by which the hand or the body [cut out by the woodcarver] was fashioned into its shape? The woodcarver will perhaps say, by the axe or the auger . . . [but] it is not enough for him to say that by the stroke of his tool this part was formed into a concavity, that into a flat surface; but he must state the reasons why he struck his blow in such a way as to effect this, and what his final object was; namely, that the piece of wood should develop eventually into this or that shape. It is plain, then, that the teaching of the old physiologists is inadequate, and that the true method is to state what the definitive characters are that distinguish the animal as a whole; to explain what it is both in substance and in form, and to deal after the same fashion with its several organs; in fact, to proceed in exactly the same way as we should do, were we giving a complete description of a couch.[17]

Such, essentially, is the doctrine of Aristotle. Assuredly, he goes too far. If he is asked what the form is which presides over the formation and functioning of the organized body, he responds: It is the soul. This notion can be discussed if one prefers; but supposing that this notion is rejected, as is frequently the case today, the facts which it sums up remain what they are. That they remain unexplained does not imply that they do not exist, and since teleology in nature is the point which interests us, we shall leave aside the notion of the soul as the possible object of another investigation. We shall be content here to say that if there is not teleology in the living being and in its relations to its surroundings, then there is no reason to suppose the existence of a soul, since, even if it existed, there would not be anything for it to explain. The notion of the soul depends upon that of teleology, not the other way around.

Another cause of confusion in discussions about teleology has to do with the introduction of the notion of "life." This is not an Aristotelian notion but a Platonic one. "Life" exists for Plato; Aristotle only knows living beings. One should not then imagine that the cause of teleology should be tied to that of "vitalism." We need not here define living beings as such. This is not our object. We only say that their correct description does not necessarily imply the appeal to a special force which is called "life." Nevertheless, it remains true that, of whatever nature the cause of the vital operations may be, it only acts in and through organized bodies endowed with a structure such that these operations are possible. The problem of natural teleology poses itself in this way. It should be resolved in its own terms, without reaching beyond it.

Assuredly, other problems may then present themselves, but only if one agrees first on the existence of natural teleology. Those who do not wish to face such problems are therefore reasonable in denying teleology. At the risk of disappointing some we shall not push the discussion simultaneously to the question of knowing whether, if there is a natural teleology, in what it consists. We know what it is in the nature of man: it is intelligence. There, where teleology is conscious of itself, it knows in what it consists; but if there is an organic teleology in animals and plants, the problem of knowing what its nature is does not allow of any demonstrable solution. It may be supposed at least that this force internal to the process in living beings is related to intelligence, whether it be that it directs itself toward intelligence as its end or descends from it as its cause. These are legitimate metaphysical speculations, and, in a sense, inevitable; but we shall restrict ourselves to the point from which one tries to determine whether they can develop and on what ground.

Let us hold, then, for the present, to the position of the problem as defined by Aristotle. To think that the perfectly regular order of the stars is the result of chance appears to him to be ridiculous, but even more ridiculous to him appears the idea that living beings might not be caused by some principle which, if it is not exactly an art like our arts, at least resembles them very much. The difference is that in the case of art the principle is exterior to the work instead of being interior to it, as it is in the work of nature. Where does this principle come from then? By what means does it penetrate matter in order to work from within? We shall ignore this, and if philosophy is free to speculate on this point, science has no obligation to speak to us about it. Aristotle only believes that for plants as for animals this immanent principle of organization can only come from without. We understand by that: from outside of matter, from which it is specifically different. "For just as human creations are the products of art, so living objects are manifestly the products of an analogous cause or principle, not external but internal, derived like the hot and the cold from the environing universe."[18]

This is as far as Aristotle is able to go. Still, we ought to assure ourselves that he could legitimately go that far.

The position of Aristotle in the matter of final causality dates from a time when the word "science" covered the totality of rational explanations based on sensible experience, which the philosophers of the thirteenth and fourteenth centuries were to call *ratio sensata*, the "sensible reason," that is to say, reason founded on sensible experience. Nothing then separated science from philosophy, for the latter was the love of wisdom or searching into and considering first principles and first causes. Each science led to the knowledge of its own principles, which constituted its own wisdom. All these particular wisdoms led to the knowledge of the absolutely first principles, which were common to them and which formed the object of the first and absolute wisdom, often called metaphysics.

Aristotle's biology is situated in this general framework. It included first of all a gathering of data positively known as a result of the work of numerous naturalists. Aristotle did not claim to be one of them, but he uses their work with abandon, and the knowledge which he owes to them he always admits deserves the respect of their successors. When modern naturalists deign to speak of this part of his work, they voluntarily recognize it as akin to theirs, indeed as among the greatest of works. He himself, however, we have seen, did not make any claim to this rank. He preferred the work which consisted in constituting the wisdom proper to each science from facts gathered by scientists.

This wisdom was in his eyes the work of reason, but since it consisted in the knowledge of the first principles of a science, and since as with other principles these are indemonstrable, Aristotle inevitably came in each science or each class of sciences to the positing of principles which are indemonstrable because primary; but they were also evidently true, because an entire order of nature becomes intelligible by their light. The notion of "end" is for him among this number. It signifies for him the limit, the achievement of growing [*devenir*] of all living beings, animal or plant. Since their development always leads to a limit which is at least provisionally felicitous, and since the reason for this success is not met with in any of their parts as parts, it is necessary that this future limit preside from the beginning over the ordering of the parts. This is what Aristotle calls the *telos, to ou eneka, to dia ti, o skopos,* or, further, the cause of the felicitous conclusion of the operation: *aitia tou eu*, that which causes the growth to take place beautifully and properly and results in a state so characterized: *to aition tou kalos kai orthos*. It does not spring from the physical order, which is that of nature (*phusis*). Perhaps it would be necessary to go beyond this order if one wished to rise to the cause of the physical cause, but this is the metaphysician's task, not the naturalist's, who in his own fashion is only a physicist. For the latter the orientation of all growth toward its end is the highest property of what he calls the "form" of the living being. This celebrated "substantial form," the nonexistence of which Descartes took upon himself to announce to the world, justifies itself in Aristotle's eyes by the sole fact that unless one assigns it as a cause, the growth of living beings becomes inexplicable from the point of view of being oriented to a limit.

There is no other reason to affirm a final cause, but this was one in the eyes of Aristotle, and we shall see that it has preserved its force in the eyes of many modern scientists, who state that those who deny natural teleology have still found nothing to explain in another way the facts which they propose to make reasonable, so they content themselves to deny it.

Protesting against such scientists, a scientist declared lately: "Finalists are perhaps right, and they have, quite surely, the right to think after their fashion; but they do not have the right to affirm that scientific evidence is on their side." These finalists do not think "after their fashion" but are constrained by the evidence of facts which in the tradition and through the example of Aristotle they desire to make intelligible. As far as I know, they do not claim anymore that "scientific" evidence is on their side; the scientific description and interpretation of ontogenesis and phylogenesis remains identically what it is without the need of going back to the first, transscientific principles of mechanism or finalism. Natural science neither destroys final causality nor establishes it. These two principles belong to the philosophy of the science of nature, to that which we have called its "wisdom." What scientists, as scientists, can do to help clarify the problem of natural teleology is not to busy themselves with it. They are the most qualified of all to keep philosophizing about it, if they so desire; but it is then necessary that they agree to philosophize.

CHAPTER II

The Mechanist Objection

ARISTOTLE FOUND TELEOLOGY so evident in nature that he asked himself how his predecessors had been able to avoid seeing it there, or, still worse, had denied its presence. He explained their error on the grounds that they were deceived on the notions of matter and substance.[1] The subsequent history of philosophy ought to confirm the correctness of his diagnosis, for insofar as the Aristotelian notion of substance as a unity of matter and form survived, the notion of teleology remained indisputable; but as early as the seventeenth century Bacon and Descartes deny the notion of substantial form (a form which constitutes a substance by its union with a given matter) and the notion of final cause becomes inconceivable.[2] In fact, substance defined by its form is the end product of generation. That which remained, once the form was excluded, was extended matter, or rather extension itself, which is the object of geometry and is susceptible only of purely mechanical modifications. Descartes consigned the entire domain of living beings, including the human body, to the realm of mechanism. The celebrated Cartesian theory of the "animaux machines," which so rightly astonished La Fontaine, perfectly illustrates this point.

Different though they may be, Bacon and Descartes had at least two points in common: their taste for knowledge that is practical and useful, and their indifference concerning philosophical notions which, though perhaps true in themselves, do nothing to increase our power over nature. Mechanism allows us to know *how* organisms function, which in turn allows us to act usefully upon them or even to fabricate similar organisms; knowledge of the final cause tells us only the *why* of mechanism, which is often obvious and permits no useful action on reality.

So far as the Greek, and even the Christian, tradition was concerned, this opinion was new. Plato, Aristotle, Plotinus, Augustine, and the long series of scholastic theologians had established as the final end of man the contemplation and love of the truth, sought out and possessed for itself. It is not only with the Gospel about Martha and Mary that the superiority of contemplation over action is proclaimed. In the *Contra gentiles* (III, 25, 15–16) Thomas Aquinas noted with satisfaction the agreement of Aristotle with Matthew and John on this point.

This view, the classic one in the great Western tradition, is admirably illustrated by what Plutarch says of Archimedes in his *Life* of Marcellus. When one reads it, for example in Amyot's translation, one comes to know what was formerly the ideal, to which the Cartesian reform wished to put an end.[3] Completely contrary to Archimedes, who nevertheless constructed more mechanical devices than Descartes even dreamed of, the author of the *Discourse on Method* saw the most certain proof of the truth of his own philosophy, and its greater merit, in its utility. Scholasticism was practically useless; therefore it was false. His own philosophy was to have practically limitless fecundity; therefore it was true. Assuredly, all true knowledge is useful, but its utility is other than the utility of machines; that a particular form of knowledge may be practically sterile does not prove that it is vacuous. But let us hear the word of Descartes himself, the new Archimedes, prophesying the coming of the age of applied science and industrialism, which in effect has come upon us:

> But as soon as I had acquired some general notions concerning Physics, and as, beginning to make use of them in various special difficulties, I observed to what point they might lead us, and how much they differ from the principles of which we have made use up to the present time, I believed that I could not keep them concealed without greatly sinning against the law which obliges us to procure, as much as in us lies, the general good of all mankind. For they caused me to see that it is possible to attain knowledge which is very useful in life, and that, instead of that speculative philosophy which is taught in the schools, we may find a practical philosophy by means of which, knowing the force and the action of fire, water, air, the stars, heavens and all other bodies that environ us, as distinctly as we know the different crafts of our artisans, we can in the same way employ them in all those uses to which they are adapted, and thus render ourselves the masters and possessors of nature.[4]

Descartes could have written *"beyond* that speculative philosophy," but he wrote *"instead* of that speculative philosophy." His philosophical reform therefore marked the revenge of Martha upon Mary, and at the same time the triumph of modern pragmatism over the contemplationism of the Greco-Christian tradition. His ambition to know the processes of nature as well as we know the tricks of our artisans – and he would have enjoyed seeing this amply satisfied today – leaves room for no doubt upon the clear-sightedness of his enterprise or about its philosophical bearing. In that Descartes did not differ from Francis Bacon, whose *New Atlantis* was more than forerunner of novels and films anticipatory of the future which flourish in our day. Bacon's work was the manifesto of the industrial age in which we live. As knowledge completely oriented toward the practical, Bacon's science was already exactly that of our time in that it postulated the primacy of action over contemplation.[5]

We are not drifting away from our problem. We are at its center, for since the efficient cause [*cause mécanique*] is the only one that gives us a grip on nature, it is the only one worth knowing. Even if there were final causality, which Descartes denied but Bacon admitted, there is no place for it in a science whose end is to make us masters and possessors of nature. Final causality by its very nature is not susceptible of being refashioned. It is superfluous to say that birds are made for flying; that is obvious. But if anyone wishes to show *how* birds fly, we would be tempted to construct some flying machines. If philosophy identifies true knowledge with useful knowledge, as modern scientism does, final causality will be by the same stroke eliminated from nature and from science as a useless fiction.

Aristotle, who was a Greek, saw things otherwise. In his philosophy final causality occupied a considerable position because its workings were, for him, an inexhaustible source of contemplation and admiration. In astronomy, in physics, and in biology he was as curious to know how things happened as our contemporaries can be, but he thought he had come across the truth of nature from the moment when he had perceived its beauty. Not so much aesthetic beauty, such as that of light and colors or forms; but first of all and above all the intelligible beauty, which consists in the apperception by the mind of the order which rules the structure of forms and presides over their relations. The order and beauty of nature interested him essentially, and not only the beauty of the heavenly bodies, these divine beings, but the harmony which appears in the structure of the most humble beings, and even in their most vile parts. The only recompense which can attend such knowledge is the joy of admiring its objects. Now, in the case of living beings there is hardly any difference between admiring the harmony which presides over their structure and discerning the teleology to which the order of their parts corresponds. The final cause is the point of view of the artist and, in the first place, of the artisan. Consequently it is the supreme object for the observer of nature to discover, he who first and foremost sets himself to contemplate beauty. There is something naive in the critique directed against Aristotle since the time of Bacon and Descartes. His philosophy is castigated for its inutility, as if the notion of a utilitarian philosophy were not foreign to his mind.

Aristotle's reasonings in favor of natural teleology appear to be extremely naive when he compares nature to an artisan fabricating a couch in metal or a bed from wood. It is in fact naive, but not without point. The consideration of the beauty of a living organism, for him who discovers the order and mutual adaptation of its parts, is as useless as the consideration of the beauty of a fine painting or a beautiful statue or, we may even say, of the beauty of a well-constructed machine. It is no less obvious, and it is always the sensible sign of a concealed intelligibility. Its inutility consists in the fact that beauty is an end it itself, not a means toward something else. No notion was more familiar to the biologist Aristotle:

> Having already treated of the celestial world, as far as our conjectures could reach, we proceed to treat of animals, without omitting, to the best of our ability, any member of the kingdom, however ignoble. For if some have no graces to charm the sense, yet even these, by disclosing to intellectual perception the artistic spirit that designed them, give immense pleasure to all who can trace links of causation, and are inclined to philosophy.

And further on he says:

> We therefore must not recoil with childish aversion from the examination of the humbler animals. Every realm of nature is marvellous: and as Heraclitus, when the strangers who came to visit him found him warming himself at the furnace in the kitchen and hesitated to go in, is reported to have bidden them not to be afraid to enter, as even in that kitchen divinities were present, so we should venture on the study of every kind of animal with-

> out distaste; for each and all will reveal to us something natural and something beautiful. Absence of haphazard and conduciveness of everything to an end are to be found in Nature's works in the highest degree, and the resultant end of her generations and combinations is a form of the beautiful.[6]

Is this attitude toward reality irrevocably passé? It is certainly not fashionable, but one can doubt whether it is foreign to the consciousness of scientists of our time. They speak of it less, but that is all. Furthermore, this is not true of all scientists. I recall the manner in which Alfred North Whitehead spoke of "the divine beauty of the equations of Lagrange." The same sentiment saw daylight [*s'est fait jour*], more recently, in an essay of Dirac's on the evolution of the image of nature in the mind of modern physicists.[7] Particularly in connection with the evolution of molecular physics, and after having recalled that the greatest modern physicists have searched for "beautiful theories," "beautiful equations," and "beautiful generalizations" in order to describe events on the atomic level, Dirac tells of the case of a physicist who had refused to give credence to his mathematical solution of a problem because it did not exactly agree with the facts of observation, even though it was the mathematical solution which was true and, as was established in the sequel, the observation which was faulty. "I think," Dirac added, "that this account bears a moral, namely that it is more important for the scientist to have beautiful equations than agreement with experience." And further, after some reflection: "It appears that if one works with the concern for having beauty in one's equations, and if the intuition one has is truly sound, one is on the right road to progress."

Pulchrum index veri! If that is true in physics, how much more ought it to be the case in biology. The word "end" no longer being popular, one prefers to speak of adaptation, but the sense is the same. In the *Origin of Species*, chapter 3, Darwin did not hesitate to write: "We see beautiful and curious adaptations everywhere in the organic world"; in chapter 4: "beautiful and curious adaptations"; and so on. The beauty of these adaptations impressed him so profoundly that he saw in them a proof of the modification of species. In the second chapter he explained without scruples – almost as naively as Aristotle – that nature could never produce at the first try organisms so marvelously adjusted and adapted to their surroundings: "Almost every part of every organic being is so beautifully related to its complex conditions of life that it seems as improbable that any part should have been suddenly produced perfect, as that a complex machine should have been invented by man in a perfect state."[8] Darwin's transformism finds in beauty something to confirm the argument for adaptations.

What motive could have incited so many moderns to eliminate from their interpretation of nature a notion still so visibly present to the minds of some among them?

If one compares the two notions of science, one would say that in eliminating the search for final causes, and in even denying their existence, Descartes took away from Aristotelian science the nature of its supreme object. Inversely, in giving himself up to the contemplation of final causes, Aristotle had retarded the birth of modern science and diverted the mechanist interpretation from the nature of its own object. This is what Bacon said with less authority but as much force and clarity as Descartes himself.

More subtle than Descartes because less systematic, Bacon never completely denied final causality (Descartes had gone so far as to deny its presence in the thought of the Creator himself); he only said that the consideration of final causes was scientifically vain. Dividing causes into two classes, physical and metaphysical, he allocated the consideration of material and formal causes to physics and that of final causes to metaphysics.[9] This separation of the physical and the metaphysical, a decision in itself quite "modern," constituted in itself a revolution of considerable importance.

The principal objection of Bacon against formal causality (in the sense of the "substantial form" or that which constitutes substances) is that it is an abstract notion, incapable as such of entering into the structure of reality. To say that man is man by virtue of the form "man" does not say anything strictly about what man is, and the same goes for the other forms. The abstraction is only necessary in order to fix the fluid contents of sensible experience. Without abstract concepts the mind would lose itself in its images of particular beings. Words signify those concepts, but it is possible to give abstract names to beings without knowing much about what they are. Searching the past for an example to cite, Bacon recalls Democritus, who, because he at least attempted to describe the structure of beings, had made more progress than Aristotle in classifying things: "But to resolve nature into abstractions is less to our purpose than to dissect her. . . . Matter rather than forms should be the object of our attention, its configurations and changes of configuration, and simple action, and law of action or motion; for forms are figments of the human mind, unless you will call those laws of action forms."[10] And why recognize in physics this superiority over metaphysics? Because, Bacon says, "physical causes shed light on new initiatives *in simili materia.*"[11] Put otherwise: physical knowledge of the material cause makes new inventions possible, while abstract knowledge of the formal cause is useless so far as practical consequences are conceived. It does not tell us how beings act, function, or live. Since they do not tell "how things work," formal concepts do not suggest any way of making machines capable of functioning and of producing in their turn other objects. Contemporary surgery illustrates remarkably this notion: an extremely exact knowledge of the actual heart and its functioning is the necessary, if not the sufficient, condition for the fabrication of artificial hearts capable of correctly taking up the function of the real ones.

After the critique of the formal cause comes that of the other metaphysical cause, the final one. With great penetration, Bacon goes right to the center of the problem. His main objection is that the contemplative enjoyment of the spectacle of final causes is what averted the attention of the ancient philosophers from the study of material and efficient [*motrices*] causes, the only ones the knowledge of which might have some practical usefulness. On this point Bacon was certainly right. Entirely absorbed by the "harmonies of nature," lost in the contemplation of their beauty, the Ancients thought they had understood nature, although they had only admired it.

It is possible to get a fairly precise idea of Bacon's point of view by comparing it to what to this day artists and writers think of the manner in which critics judge their work. The critic hastens to say if a work is beautiful or ugly, that is to say, if it pleases him or not. He says next, most frequently, what it represents or signifies. Finally, if he takes the trouble, he tells us whether the parts of the work, its materials and forms, appear to him to be well proportioned to each other and adapted to their purpose. In brief, in the most favorable cases the critic strives to say *why* the artist made his work such as he did, but to say *how* he made it would be a completely different affair. This is indeed why artists are often in disaccord with their critics, because for the artist the problem is not to make something beautiful, but to find out how to make it. In the arts relating to what is pleasing still more than in nature, beauty is the end, not the means. One can become aflame with the desire to emulate beauty while enjoying the pleasure of contemplating it, but that offers no instruction on how to do so. The art of creating something beautiful is not learned by admiring it where it exists, but rather by searching out the ways followed by nature and art in creating it.

Bacon deserves our complete attention on this point, for he is right. He does not say that there are no final causes; he simply says that their study has been *misplaced*, that we are deceived about the place which belongs to it. Here, once more, we can do no better than to cite him:

> But this misplacing hath caused a deficience, or at least a great improficience in the sciences themselves. For the handling of final causes, mixed with the rest in physical inquiries, hath intercepted the severe and diligent inquiry of all real and physical causes, and given men the occasion to stay upon these satisfactory and specious causes, to the great arrest and prejudice of farther discovery.[12]

Let us repeat it: Bacon is right if practical utility is taken as the criterion of philosophical, or even scientific, truth. But if there is really final causality in nature, it is then still necessary to take it into consideration. And if to know it involves as a consequence the admiration of its beauty, the contemplation of natural beings, which no naturalist in fact fails to do, makes an integral part of the knowledge which we have of them. It is true, and Bacon in fact justly notes it, that an excessive naivete often mars the search for final causes. He brought into disrepute in advance the future "simplisms" of Bernardin de Saint-Pierre: "For to say that eyelashes are there to provide a quickset hedge and protection about one's view; or that the firmness of the skin and flesh of animals is there to protect them against the excess of heat and cold . . .; or that the leaves of plants are there to protect the fruit;[13] or that the clouds are there to diffuse the sun, and so on," all that may be proper in metaphysics, but it is out of place in physics. Similar to the lamprey [*remora*] clinging to the side of a vessel and impeding its progress, the search for final causes has had the effect of retarding the search for physical causes.[14] In this case Bacon's judgment is that of the history of the sciences; it allows of no appeal. The contemplation of nature and its beauty has certainly retarded scientific research into nature's properly physical structure. Scientists are resolved that this error will not be repeated, and the violence of their attacks against final causality are explicable at least in part by that consideration. If that fear were not henceforth superfluous, one could even call it justified.

It is, however, superfluous, because nothing prevents the two points of view from coexisting, and if their peaceful coexistence is possible, it is desirable. A half-truth is never worth a whole truth, and, in fact, these two parts of the truth have coexisted, even after Bacon, in scientific minds far superior to his, and even after Descartes, in geniuses who were certainly not inferior to him.

It might not be appropriate to call the eighteenth century the metaphysical century, but the fecund interest it bore for the sciences did not prevent it from enjoying the contemplation of the "harmonies of nature." There is no reason why we should deprive ourselves of such contemplation. First, if recourse to final causes is forbidden because their role is unintelligible, one would have to forbid recourse to the cause called efficient, or simply to the cause of motion, for the same reason. In what sense is efficient and mechanical causality more intelligible than final causality? It was so in the Aristotelian world of the formal cause, especially since the existential act of being had been posited by Thomas Aquinas as the act of acts and the perfection of perfections. But Malebranche, and Hume after him, established the fact that the problem of the "communication of substances" becomes insoluble in a universe deprived of all substantial form and completely mechanized. Noting that state of affairs, Comte later came to conclude that, the notion of cause being unintelligible, science ought henceforth content itself with formulating laws. But if Comte were right, there is no causality of any sort in nature, and, consequently, no scientific question can be posed on this subject. We say that he ought not to have posed any. Things stand today as they did in Aristotle's time: living beings continue to be composed of heterogeneous parts ordered according to determined relations, and the order of the mutually adapted parts remains today as then inexplicable in terms of the efficient or motor cause alone which moves matter exclusively according to the laws of the mechanics of solids, liquids, or gases. A harmony in fact exists, whatever its nature may be, between the heterogeneous parts of an organism, just as a harmony exists between the parts of a machine. In brief, if there is in nature at least an apparently colossal proportion of finality, by what right do we not take it into account in an objective description of reality?

It is there, let us recall, that, according to Aristotle, the heart of the matter lies. If the scientist refuses to include final causality in his interpretation of nature, all is in order; his interpretation of nature will be incomplete, not false. On the contrary, if he denies that there is final causality in nature, he is being arbitrary. To hold final causality to be beyond science is one thing; to put it beyond nature is something completely different. In the name of what *scientific* principle could one exclude from a description of reality an aspect of nature so evident? Explanations which rely on final causality have often been ridiculed, but mechanist explanations have often been ridiculed also, and this does not disqualify the legitimacy of either point of view. The impressive declaration of Aristotle's in the first chapter of *On the Parts of Animals* ought never be forgotten: "But if men and animals and their several parts are natural phenomena, then the natural philosopher must take into consideration not merely the ultimate substance of which they are made [today we would say their physicochemical elements] but also flesh, bone, blood," in addition to the heterogeneous parts such as the face, hands, feet; he ought to search out "how each of these comes to be what it is, and in virtue of what force." In sum, since animals have at one and the same time form and structure, their "shape and structure must be included in our description of them." Aristotle goes as far as to say that the consideration of the formal cause is more important than the consideration of the material cause,[15] which is debatable because he who loses himself in the contemplation of the form opens himself to the possibility of allowing many a secret to remain hidden in unexplored nature. But it is possible to take account of one without excluding the other, and that is all that we wish to point out.

However that may be, the efficient (or motor) cause and the material cause evidently deserve more esteem than Aristotle accorded them. We live in the age of Descartes and Bacon, and the colossal success of the applied sciences in industry is proof of it. The date of their first victory is unknown. Robert Lenoble, priest of the Oratory, entitled his work on one of Descartes' friends *Mersenne ou la naissance du mécanisme.*[16] From the first pages of his work the author observed that whoever approaches the seventeenth century by coming, as he ought, from the philosophy of the sixteenth century sees springing up before and around Descartes many currents which go to form modern thought. They all have one common characteristic: mechanism.

The first great and indisputable triumph of mechanism was

the astronomy of Newton. However, Newton gave proof of more prudence than Bacon and Descartes in his philosophy of nature. Further, in 1704, in his *Opticks*, although he found himself at grips with the strictly mechanical physics of the Cartesians, according to whom all luminous phenoma ought to be caused and propagated by pressure and movement, and seeing that the theory of Huygens did not agree with the facts, he himself fell back upon his own favorite theory of an ether serving as the milieu for the propagation of luminous rays. He gave as proof of it something which appears today to be a curious process of scientific reasoning. Speaking of those who denied his theory of a gravitational force, he reproached recent philosophers for banishing "the consideration of such a cause out of natural philosophy, feigning hypotheses for explaining all things mechanically, and referring other causes to metaphysics; whereas the main business of natural philosophy is to argue from phenomena without feigning hypotheses, and to deduce causes from effects, till we come to the very first cause, which certainly is not mechanical."[17]

There follows then in Newton's text a long series of questions that mechanist science leaves without answer, or in view of which, in order to find answers to them, the mechanists invent gratuitous explanations. Provided that their answers are nothing but mechanical, their inventors judge them true, but Newton is not convinced. What an extraordinary reversal of the situation created by Bacon! The same Newton who said "I make no hypotheses" (*hypotheses non fingo*) here rejects mechanist hypotheses in order not to prevent science from considering questions to which the discovery of mechanist answers is improbable. Among these questions there are several that Aristotle would have had pleasure in coming across: "Whence is it that nature doth nothing in vain; and whence arises all that order and beauty which we see in the world? . . . How came the bodies of animals to be contrived with so much art, and for what ends were their several parts? Was the eye contrived without skill in Optics, and the ear without knowledge of sounds? How do the motions of the body follow from the will, and whence is the instinct of animals?"[18] But since our own reflection before all else bears on final causes in biology, let us consult on this precise point a biologist of the nineteenth century, Claude Bernard.

It is possible to take him as representative of the spirit of scientific research in its purity. Having remarked that among the anatomical pieces laid on the table the flies prefer liver, he concluded that they searched out the sugar, but he asked himself immediately how it was that sugar came to be located there. His initial supposition was naturally that the liver stored up the sugar contained in the food, but experience obliged him to conclude that the liver did not borrow the sugar it contained. It produced it. By means of a capital generalization he then inferred that animals themselves are capable of carrying out directly the organic synthesis of vital elements. By a still more bold generalization he inferred that the same power ought to be attributed to vegetables themselves: if there is sugar in certain plants, then they ought to be the ones who make it. The first stage of this vast generalization is represented by Claude Bernard's doctoral thesis: "Recherche sur une nouvelle fonction du foie considéré comme un organe producteur du sucre chez l'homme et chez les animaux," March 17, 1853. The second stage is represented by a course of lectures recently reedited and republished under the title of *Leçons sur les phénomènes de la vie communs aux animaux et aux végétaux.*[19] This last work contained the results of a life consecrated to a scientific research which did not exclude authentic philosophical reflection.

It might be useful to note that Bernard was not a vitalist in biology. When asked for his opinion about life, he replied that he had none, never having encountered life. On the contrary, the observable vital properties of matter in plants and animals were in his eyes undeniable facts. He summed these up under five categories: organization, generation, nutrition, growth, and, finally, decay, ended by sickness and death. He made no appeal to any "life" in order to explain these functions. He had recourse to no "soul" to explain the presence of these activities in living beings. He was so inflexible on this point that he even refused to hold organization as a principle. It was not, in his eyes, a power of organizing, but a fact. Another biologist, named Rostan, had placed in organization the fundamental character of life, which had given to his doctrine the name of *organicism.* Rostan was not a vitalist either. He even refused to think of organization as a force added to organized being. On the contrary, he defined it as the power "which results from structure," not, consequently, a property distinct from the machine or a quality superadded, but, simply, the machine itself once assembled. Briefly, Rostan said, "organization is the machine assembled."[20]

It is remarkable that this apparently radical mechanism did not satisfy Claude Bernard. What discomfited him in the organicism of Rostan was what today would make of it a success with philosophers, namely, its "structuralism." Why should life come about from a "structure"? Structure, Claude Bernard observed, is a vague notion: "it is not a separate property";[21] it is not a force capable of bringing about whatever a thing may be; otherwise it would itself need to be explained by another cause. How, then, will Claude Bernard define organization? He will not. He gives up trying to "define the indefinable" and will content himself with characterizing living beings by opposition to inorganic matter. Organization, he says, "results from a blending of complex substances reacting upon each other. So far as we are concerned, it is the arrangement which gives birth to immanent properties of living matter. This arrangement is quite special, and quite complex, but nevertheless it obeys the general chemical laws of the arrangement of matter." And then the conclusion: "Vital properties are in reality only the physicochemical properties of organized matter."[22]

This is perfectly clear. The final cause could not be found in living beings such as Claude Bernard conceives them. But there is a flaw in this metal. Bernard says: "The physicochemical properties of organized matter," without saying from whence this organization comes to matter. We have understood him to say that organization results from the blending of complex substances reacting upon each other. The problem remains: How has their blending become an organization?

This was the Achilles' heel of the theory, and Claude Bernard knew it. Everything in living organisms is physicochemical. Our laboratories can reproduce synthetically, for example, fats naturally present in organisms. But, in the first place, the processes of production are not the same;[23] and, further, in examining the series which goes from living protoplasm to organized living beings, Bernard made the observation that protoplasm itself is not yet a living *being*. It could be a living thing; it is not yet a being. Protoplasm is living matter; in order to become a living being it requires a form.[24] "The form of life is independent of the essential *agent* of life, the protoplasm, since the latter remains such throughout infinite morphological changes."[25] The same question then occurs again: How are we to give an explanation of these forms, figures, structures, call them what you will, which are the very natures of animals and plants and preside over their development?

Two points should be noted in Bernard's response. First, the final cause has no place in science because "the final cause does not intervene as an actual and efficacious law of nature."[26] This amounts to saying that the final cause is not efficient, and since science is uniquely interested in the efficient cause, it does not have to take the final cause into consideration. But that is not to say that there can be no teleology in nature. The Aristotelian position on the problem is then intact. But if teleology is not a law of nature,[27] what is it?

At a certain point Bernard replies that, rather than a law of nature, final causality is "a rational law of the mind [*esprit*]."[28] That sounds a bit like Kant, but how could a law of the mind explain the organization given factually in actually existing beings? Bernard adds that determinism is "the only possible scientific philosophy."[29] But teleology does not propose to eliminate determinism; it proposes to explain the existence of the mechanically determined. In fact, the refusal to involve final causality in order to explain organization in nature amounts to leaving unexplained the very existence of organisms. And Claude Bernard knows it: "The general agents of physical nature capable of causing the appearance of isolated vital phenomena do not explain the general ordering, the *consensus* and concatenation of it."[30] And nevertheless this consensus exists in nature. Bernard describes the biological function as "a series of acts or of phenomena grouped or harmonized in view of a definite result. . . . These constituent activities proceed one through another; they are harmonized or concerted in a fashion so as to come together in a common result."[31] We in our turn may then ask what is this result *in view of which* the acts of a series are grouped if it is not their final cause, their end?

Thus, just as beauty remains for certain modern scientists an indication of truth, the simple fact that organized bodies exist invites still other modern biologists to look in nature for a principle which presides over the organization of living beings. Without such a principle the functioning of such beings can be explained, but not their existence, which, after all, is as much a fact as is their functioning. An adversary of final causality under all its forms, our contemporary Jean Rostand, concludes on this point: "We ought to admit that organic adaptation, in its entirety, still awaits its exhaustive explanation."[32]

And wait it does! A few centuries more or less will not make a big difference. If it is true, as we think, that scientism looks for the explanation in the wrong direction, it will only cause its response to be postponed longer and longer. Meanwhile, grappling with the problem philosophically, we ought to feel ourselves free to ask if there is not, in the very nature of things, a reason why a *scientific* solution to the problem is essentially impossible.

CHAPTER III

Finality and Evolution

ARGUMENTS FOR AND AGAINST final causality hardly changed from Aristotle to the beginning of the nineteenth century, which saw the advent of transformism and of the notion of biological evolution. Initially it will not be useless to cast a glance at the doctrine which evolutionism discomfited.

A. Fixism

Today fixism is the name given to the view of the world which is opposed to transformism. It was held as so obviously evident that it was not felt useful to designate it with a particular name. For the same reason it did not appear needful to define it. In this respect it could be said that it is transformism which created fixism, the latter only defining itself with any precision when possibilities of doubting it—indeed sometimes temptations to doubt it—present themselves to the thought of those who speak about it. It is, however, certain that the traditional teaching of Christian theology was an invitation to conceive of the world as being at present such as it had been since its creation. In accordance with the requirements of theological method, which moves from God to things, what ought to be the nature of things was deduced from the nature of God. A divine and unchanging cause could only have created definitively.

It appears that the problem was posed with perfect clarity for the first time in the mind of Descartes. When he had to explain the structure of the world in his *Principles of Philosophy*, he found that the philosopher in him came to grips with the Christian. As philosopher, he ought naturally to follow in his account the order of the generation of things, from the simple to the complex; as

Christian he could only defer to the authority of revelation, or, what practically came to the same thing, to what he thought had been revealed by God.

In substance that had already been the attitude of Thomas Aquinas. Moreover, for purely philosophical reasons, he took the view that beings had been created in their perfect state: *naturali ordine perfectum praecedit imperfectum, sicut actus potentiam.*[1] On the other hand, and inversely, if one moves from the order of creation to the order of natural generation, the latter always proceeds from the imperfect to the perfect: *natura procedit ab imperfecto ad perfectum in omnibus generatis.*[2] When, therefore, revelation teaches something on the subject of creation, it is necessary to accept such teaching as true; but for the rest, reason ought to be followed: *Unde, in omnibus asserendis, sequi debemus naturam rerum, praeter ea quae auctoritate divina traduntur, quae sunt supra naturam.*[3]

Thomas Aquinas thought that God created living beings at maturity, since he created them with the perpetuity of the species in mind, and thus capable of reproducing themselves (*S.T.* I, 94, 3). Descartes set out from another theological principle but drew from his own theology of the infinite God the conclusion that one could not think too highly of his works. Setting out from the notion that one ought not fear deceiving oneself by imagining the works of God to be too beautiful, too great, or too perfect, he comes back again to the conclusion of Thomas Aquinas:

> I have no doubt that the world was created in the beginning with such perfection as it possesses, in such fashion that the sun, the earth, the moon, and the stars have existed from that time. And the earth not only had in it the seeds of plants, but indeed the plants themselves covered a part of it; and Adam and Eve were not created as infants but as mature human beings. The Christian religion wills that we understand things thus, and natural reason completely convinces us of this truth. For if we consider the omnipotence of God, we ought to judge that all that He has made has had from the beginning all the perfection that it ought to have.

Let us retain these words: "and natural reason completely convinces us of this truth," for we shall find them again before long taken in a purely scientific context. It turns out merely that Descartes imagined (what Voltaire will call Descartes' fiction) a possible explanation of the entire universe, including living beings, from simple material elements, without having recourse to any form or, properly understood, to any teleology. It was to be, in brief, a purely mechanist and yet genetic explication of the universe.

> But nevertheless, we come to know much better the nature of Adam and that of the trees of Paradise if we have examined how children are formed bit by bit in the wombs of their mothers and how plants spring from their seeds, than if we have only considered what they were when God created them. Likewise, we shall better understand what is generally the nature of all the things which are in the world if we can imagine several principles which are quite intelligible and quite simple. By such principles, then, we might be able to see clearly that the stars and the earth and at length the entire visible world could have been produced, as it were, from several seeds (though we know that it was not produced in that manner), if we describe it only as it is rather than [*ou bien comme*] as we believe that it was created. And because I think that I have found such principles, I shall try to explicate them here.[4]

Descartes found himself then in a situation analogous to that of the Latin Averroists of the thirteenth century. He has two different explanations of the same facts: one which he pretends to believe or believes because he is a Christian; the other which he likes because it is pleasing to his reason. And he maintains both of them. Only, in order that he might not be accused of teaching the doctrine of "the double truth" formerly condemned in theology, he goes further than any known Averroist had ever gone and declares that the philosophical conclusion which he proposes is neither necessary nor true. "And so much is it not the case I wish others to believe everything that I shall write, that I even here propose some things which I believe to be absolutely false, namely: I do not doubt . . .," and so on. In whatever fashion the doctrine may have come to Darwin's knowledge, through intermediaries whose names are unknown to us, it is the decay of this same notion in his mind which will determine his passage from fixism to transformism. He will be persuaded that the Christian religion teaches the creation of beings such as we know them at the present time, and when his own observations and reflections render this belief impossible to him, he will lose his initial faith in the truth of the Christian religion.

Between Descartes and Darwin Linnaeus takes his place, whose fixism does not serve as an excuse for any covert evolutionism. His *System of Nature* is the work of a classifier who first set himself the task of reducing to a scheme the three kingdoms of nature: the mineral, the vegetable, and the animal. Aristotle, who would have seen in Descartes a new Empedocles to combat, would probably have found nothing blameworthy in Linnaeus.

The work begins with an invocation to the Creator:

> O Lord, how manifold are thy Works!
> in wisdom thou hast made them all:
> the earth is full of thy riches
>
> Psalm 104:24

There follows the first table, entitled:

Observations on the Three Kingdoms of Nature

1. If we observe God's works, it becomes more than sufficiently evident to everybody that each living being is propagated from an egg and that every egg produces an offspring closely resembling the parent. Hence no new species are produced nowadays.

2. Individuals multiply by generation. Hence at present the number of individuals in each species is greater than it was at first.

3. If we count backwards this multiplication of individuals in each species, in the same way as we have multiplied forward (2) [sic], the series ends up in one single *parent*, whether that parent consists of *one single* hermaphrodite (as commonly in plants) or of a double, viz. a male and a female (as in most animals).

4. As there are no new species (1); as like always gives birth to like (2); as one in each species was at the beginning of the progeny (3) it is necessary to attribute this progenitorial unity to some Omnipotent and Omniscient Being, namely *God*, whose work is called *Creation*. This is confirmed by the mechanism, the laws, principles, constitutions, and sensations in every living individual.

* * * * *

8. Natural objects (7) belong more to the field of the senses (5) than all the others (6) and are obvious to our senses anywhere.

Thus I wonder why the Creator put man, who is thus provided with senses (5) and intellect, on the earth globe, where nothing met his senses but natural objects, constructed by means of such an admirable and amazing mechanism.

Surely for no other reason than that the observer of the wonderful work might admire and praise its Maker.

* * * * *

10. The first step in wisdom is to know the things themselves. . . .

11. Those of our scientists, who cannot class the variations in the right species, the species in the natural genera, the genera in families, and yet constitute themselves doctors of this science, deceive others and themselves. For all those who really laid the foundation to natural science have had to keep this in mind.

* * * * *

14. Natural bodies are divided into *three kingdoms* of nature: viz. the mineral, vegetable, and animal kingdoms.

15. *Minerals* grow; *Plants* grow and live; *Animals* grow, live, and have feeling. Thus the limits between the three kingdoms are constituted.[5]

The first impression which reading this occasions is that of a feeling for nature which is intensely religious. The rules of classification, the very idea of natural science, are not separated here from the great principles of natural theology passed on by tradition. The form of the text is no less striking, quasi-geometrical or Spinozist, with its references to one definition or another and–finally, and for the biologist primarily–with its initial assertion that, since the creation, no new species have been produced. But the philosopher cannot help but note another hackneyed thesis concerning biology and natural philosophy: all species run back to a first member or to a first couple. Finally, perhaps it would be fitting to note once more that, far from excluding mechanism, Linnaeus' finalism requires it. If living beings have been willed into existence in order to arouse admiration in the mind of the spectator, and the adoration of their author, nothing could serve this end better than knowing their mechanism. Once more, then, the close alliance of finalism and mechanism is here confirmed.

From the point of view of the modern history of biology, however, it is the third proposition which is without doubt the most important: every present vegetable or animal series goes back to an initial ancestor or, as it may be, to a first couple, male and female, from which it descends. We find the same thesis reaffirmed, still more energetically, if that is possible, in the *Fundamenta botanica*.

The issue was of great importance in Linnaeus' eyes, for the fixity of species since the creation was for him a condition of the very possibility of natural science. The ancient Greek notion, at least the Platonic and Aristotelian notion, that there is no science except of the necessary appeared to haunt Linnaeus' mind. If species vary, that is an end to classifications, and classifying living beings is biology itself. "*Botanica innititur fixis generibus.*"[6] These fixed genera exist if all living beings descend in a regular way from some ancestor or original couple. "Reason suggests the thought that at the beginning of things, a unique couple was created for each species of living beings."[7]

Linnaeus is one of the first witnesses to this thesis who would exercise a considerable influence on the history of zoology. Let us note well, however, that under this form the proposition only appeals to reason (*suadet ratio*), not to revelation. Linnaeus does not say that we ought to hold the belief, as a verifiable revelation, that God first created for each species a single couple, or, moreover, that species were perpetuated, always identically the same, from the day of creation. He is persuaded of the fixist position, for otherwise botany and zoology would have the solidity of their foundations compromised, but he does not make this a truth of faith.

It could be that theologians before or after Linnaeus have done what he himself appears not to have wished to do. Whatever the case may be, it is curious to find the same thesis reaffirmed once more, and this time as a truth which it is necessary to believe, by a naturalist whom nothing obliged to undertake such a responsibility. In his well-known chapter on the ass, after having compared it to the horse in all possible respects, Buffon comes to the conclusion that seeing such striking analogies, one would readily hold the ass to be not a really distinct species from that of the horse, but rather a degenerate horse. As he recovers his balance in order to pursue this quite seductive idea, Buffon firmly declares:

> But no, it is certain, through revelation, that all animals have equally participated in the gift of creation; that the two first

> members of each species, and of all the species, have come forth completely formed from the hands of the Creator, and we ought to believe that they were such, more or less, at creation as they today present themselves to us in their descendants.[8]

Here indeed is bad theology, but there is no doubt that Buffon did not invent it. Creation is not a gift [*grace*], since before it there was no nature to receive it. But let us move on! To know the history of popular or common theology would be, in the event, more useful than to be acquainted with the authentic teaching of the theologians. We only pose the problem here because about 1850 Darwin would find himself at grips with this same thesis. Buffon, after having been brought to believe in it, went on to convince others of it in his turn. We shall see what a decisive role this thesis played in the history of Darwin's thought.

On the other hand, less obviously but yet really, Buffon worked in novel ways. He detested classification, classifiers, and, above all, Linnaeus. The notion that natural beings form a continuous and statically ordered hierarchy, like that of an army, was already familiar to Aristotle, but he did not conclude from it that the simplest were the ancestors of the others. His taxonomy was not a genealogy. Aristotle found it easy to define the genera and the subgenera in logic; it sufficed, to apply his theory, to choose some favorable example. Thus, the species "man" is distinguished from the genus animal by his "difference," namely, "reasonable." We know that Aristotle found grave difficulties in defining and classifying the natural species through this system. The difficulty never ceased to exist, but Buffon assigned a cause to it. Doubtfully and hesitatingly he came to conclude that, strictly speaking, there are no precisely defined species. There are species, but with all sorts of passages from one to another which confer on this hierarchy a kind of continuity:

> Nature proceeds by unknown gradations and consequently does not lend herself totally to these divisions, since she moves from one species to another, and often from one genus to another, by imperceptible nuances. So that we find a great number of "average" species and objects equally divided which we know not where to place, and which necessarily disrupt the project of a general system.[9]

When Buffon follows his momentary mood, he goes to extremes. He does so in this case. In fact, if one continues to speak in this fashion, one will say that "the more one augments the number of divisions of the production of nature, the more one approaches the truth, since there are really only individuals in nature, and genera, orders, and classes only exist in our imagination."[10] This is hastily said, but our naturalist encounters here one of the most ancient constants in the philosophy of nature, a constant whose sense philosophy has never succeeded in clarifying. Aristotle already thought that only individuals exist, and therefore there ought not be species; yet they exist. There are species which, such as they are, appear to be quite real, but which, since individual substances alone are real, do not exist. This is the celebrated problem of the universal, and it is fashionable to make fun of the Middle Ages for having reduced philosophy to this problem. But the Middle Ages only said that all the rest of philosophy depended on the response made to this problem, which is the case. The modern response presupposes the negation of the notion of "substantial form," which ought logically to entail in effect the negation of species, and it does deny them, but it unscrupulously calls them back each time it has need of them; and the only means of getting beyond the issue is to deny absolutely the legitimacy of all classification. This agrees poorly with common sense, but petrography, mineralogy, botany, and zoology agree with it no better. How could one find intermediaries between classes if the notion of classes corresponded to nothing real?

Science can come to an agreement in this matter with an ease which surprises the philosopher. Buffon speaks incessantly of nature,[11] but he has some difficulty in specifying what she is. At one time he speaks of her as an assemblage of laws, at another as a being, or as a force analogous to that which Alan of Lille, in the twelfth century, called the servant of God. Starting from this ill-defined notion, Buffon proceeds to the *Views* of nature. One should not be astonished, therefore, if he does not always see the same thing, but he is aware of the fact. Furthermore, it has doubtless been noted that in denying the existence of orders, genera, and classes, which exist "only in our imagination," he did not make mention of species. Is this intentional? One would think so at first, for if individuals alone exist, in what sense could species exist? But in the "second view" of the universe there is no doubt about his intention. Not only do species exist, but they alone exist. The species is everything; the individual nothing.[12] We have therefore returned to Linnaeus, and not only does classification become possible again, but science can have no other object.

These oscillations are not at all scandalous; they are inscribed in the very nature of the problem of universals: it is true that species do not exist; it is also true that no individual exists outside of a species. Buffon therefore speaks at one time like Aristotle, at another like Plato, and he takes the two sides, not knowing which to choose. It is a very old history:

> Assidet Boetius stupens de hac lite
> Audiens quid hic et hic asserat perite,
> Et quid cui faveat non discernit rite,
> Nec praesumit solvere litem definite.[13]

Buffon nevertheless has reason to spare species in this massacre of the universals, and it is fitting to note it, because the problem which species presents–to him also–is a constant of natural philosophy. Since the time of Aristotle, and still more since Buffon's time, there has been reason to intercalate a sort of division between the highest genera and individuals. In dividing the orders into families, genera, and classes one comes to groups of living beings whose coupling is sterile. They do not reproduce. These are "mules," the typical example of which is the hybrid of the male ass and the female horse; but there exist many other cases of this in zoology and in botany. We mention this well-known fact here because in our own inquiry we come across it for the first time with Buffon, who sees himself constrained to profess a certain fixism at the point where he comes across a couple incapable of reproducing itself according to the law of heredity.[14]

These views might lead Buffon rather far.[15] Of the classifications which he discouraged he took exception to none so firmly as he did to "families." In effect, if the word has a precise sense, it is that of lineage, a group whose members are united by lines of descent from one common stock. On the one hand, he insists on the idea that in the same way that "species is only an abstract and general word," so also "we must not forget that *families* are our work; that we have made them only for the solace of our mind; that if it is impossible to comprehend the real series of all beings, it is our fault and not nature's, which only contains individuals." If this is so, however, how does it come about that, descending the chain [of being] by degrees, one comes to individuals whose fecund interbreeding is impossible? And why, ascertaining that things are thus, does Buffon turn back against the notion of "family," as if he had not established the vanity of it? It is because he takes the word seriously. If we admit that the connections between species can be of a familial type, then any species can be descended from any other, and Buffon is brought to a standstill upon the threshold of this universal transformism, hypothetical in his mind, but menacing:

> If one once admits that there may be families among plants, and among animals, that the ass may be of the family of the horse and that it differs from the horse only by having degenerated, one could say equally that the ape is of the family of man, that he is a degenerate man, that man and the ape have a common origin, just as the horse and the ass; that each family, equally among animals as among vegetables, has only had one single stock; and even that all animals have issued from only one single animal which, over the course of time, has produced, through perfecting itself and through degeneration, all other races of animals. Naturalists who so lightly establish "families" among animals and vegetables do not appear to be sensitive to the complete extent of these consequences, which reduce the immediate product of creation to as small a number of individuals as one [might] wish.[16]

That before which Buffon recoils in this remarkable passage is the future transformism of Darwin with its ineluctable consequence, *The Descent of Man.* The step he hesitated to take then had to be taken finally.

B. Transformism

We shall understand by transformism every doctrine which affirms that animal or vegetable species have changed in the course of time, no matter how these changes are explained. Transformism perhaps is better defined, in a negative fashion, as the negation of "fixism," namely, as stating that it is not true that species are today that which they were at their origin, no matter how one might conceive that origin.

1. Lamarck

Transformism is ordinarily associated with two names: Lamarck and Darwin.

J.-B. de Monet, Chevalier de Lamarck, born in 1744 in Bazentin, died in Paris in 1829, is a naturalist whose person and career defy imagination. From the point of view that interests us, his masterwork is the *Philosophie zoologique.*[1] This title itself gives notice of the nature of the work. It belongs to a time when a scientist [*savant*] did not fear that he would bring his scientific work into disrepute by presenting it as philosophy. But it is necessary to recognize at the same time that this work is presented under an aspect frankly other than that of a scientific work of the nineteenth century. It is hard to imagine Darwin presenting himself as a philosopher, and nothing resembles less the sobriety of this friend of facts than the expansiveness of Lamarck, who is always taking to reason and argument.[2]

Nevertheless, Lamarck took a decisive step, the very one that Buffon did not wish to take. He knew Buffon quite well, for Buffon had interested himself in him at the beginning of a career characterized by poverty and difficulty. Whatever the form of his demonstrations might be, the view of nature which they set forth differs significantly from that of Buffon. They are in opposition to one another, rather than differing, and the opposition is striking.

It is impossible to take a better overview of Lamarck's review of nature than the one he himself gives in the table of contents of his *Philosophie zoologique*, second part, chapter 6: "That, since all living bodies are productions of nature, she must herself have organized the simplest of such bodies, endowed them directly with life, and with the faculties peculiar to living bodies. – That by means of these direct generations formed at the beginning both of the animal and vegetable scales, nature has ultimately conferred existence on all other living bodies in turn."[3]

The presence in the passage of the *scale* [*échelle*] of beings will be noticed, a constant universally present since Aristotle; but it is especially important to note the presence of "direct generations" [Elliot. Lamarck has *"générations spontanées"*], the only nonmetaphysical, nontheological response to the question of the true initial origin of species. For Lamarck the question is one of knowing how, given these primitive organisms, the more complex vegetable and animal organisms can be formed "progressively."

The very possibility of the question presupposes the abandonment of the ancient belief in the fixity of species. One sees how resolutely Lamarck does so in reading the third chapter of the first part of his work: "Of Species among Living Bodies, and the Idea That We Should Attach to That Word." With Lamarck we arrive at a generation which is conscious of the identity of the problem: before saying whether species change and how they change, it is necessary to know what we call a species. Unfortunately Lamarck hardly went beyond the point where his predecessors left the question: "Any collection of like individuals which were produced by others similar to them is called a species." He immediately proceeds: "but to this definition is added the allegation that the individuals composing a species never vary in their specific characters, and consequently that species have an absolute constancy in nature. It is just this allegation that I propose to attack, since clear proofs drawn from observation show that it is ill-founded."[4] This conclusion entails another, which today appears to be empty of scientific interest, but of which it is important to take notice because it will come to play a decisive role in Darwin's reflections: contrary to what Linnaeus sustained in the name of reason, and Buffon in the name of revelation, there is no reason to think that each species had been the object of a "particular creation"[5] on the part of God. The problem then is one of knowing how actual species are constituted.

To begin with, Lamarck reaffirms, with at least as much decision as Buffon, that species have no real existence in nature. With the happy philosophic casualness of authentically scientific minds, he declares that all that exist are individuals who succeed one another [in time] "and resemble those from which they spring."[6] This resemblance leads to the formation of collective images of certain groups of similar individuals and, by this same means, to the notion of species, genus, or class. From the moment that the problem of the possibility of such groups is posed, we are in the presence once more of the problem of universals. One is irresistibly reminded that Lamarck reproaches Linnaeus, Buffon, and other "classifiers" especially with introducing an artificial order in nature. He himself wants us to study "the natural method," that is to say, that "our classifications should conform to the exact order found in nature, for that order is the only one which remains stable, independent of arbitrary opinion, and worthy of the attention of the naturalist."[7] A naturalist persuaded of the reality of species could say it no better.

Now, it is precisely these same species, upon which a stable order is founded, which have given evidence in the course of the ages of a certain instability. Species have "in the course of time changed their characters and shape."[8] Not only that, but our perceptions [*prises*] being such as they are, they only distinguish imperfectly one entity from another. It is difficult to determine species, and still more difficult to determine genera.[9] We observe, under the name of species, provisionally stationary states between two mutations. This stability moreover is tied to that of their conditions of existence, so much so that if the conditions of existence do not change sensibly, then the species is not subjected to any cause of variation.[10] When, on the contrary, the environment changes, then living beings change in order to adapt to the changes, as happens obviously enough for us to see the changes undergone by the same plant, or the same tree, according to the various altitudes at which we observe them. This fact leads to a new definition of species. By this name will be designated "any collection of like individuals perpetuated by reproduction without change, so long as their environment does not alter enough to cause variations in their habits, character and shape."[11] It remains to be ascertained how circumstances act on living organisms.

In fact, Lamarck comes to specify it: variations in the surroundings are the cause of the modifications in the habits of organisms. This notion of habit is of great importance in Lamarck's doctrine. It is habit which explains the reaction by which the living animal or plant undergoes changes of form in order to adapt itself to novel situations in which it finds itself placed.

No other part of his doctrine has undergone more severe criticism than this, which is understandable, since it is the key to the other parts. A similar criticism will attend the Darwinian doctrine of natural selection, which is the key to his own brand of transformism. In saying that "the environment affects the shape and organization of animals," Lamarck does not mean that the environment acts directly on the organism, but that it forces the organism to modify itself in order to adapt to the new surroundings. To speak summarily, but not inexactly: "great alterations in the environment of animals lead to great alterations in their needs, and these alterations in their needs necessarily lead to others in their activities. Now, if the new needs become permanent, the animals then adopt new habits which last as long as the needs that evoked them." On the basis of which Lamarck concludes, apparently satisfied: "This is easy to demonstrate, and indeed requires no amplification."[12]

The articulation of the doctrine is located at a precise point, which is the connection between need and habit: "Every new need, necessitating new activities for its satisfaction, requires the animal, either to make more frequent use of some of its parts which it previously used less, and thus greatly to develop and enlarge them; or else to make use of entirely new parts, to which the needs have imperceptibly given birth by efforts of its inner feeling; this I shall shortly prove by means of known facts." We add to this that these acquired modifications are transmitted by heredity: "All the acquisitions or losses wrought by nature on individuals, through the influence of the environment . . . all these are preserved by reproduction to the new individuals which arise, provided that the acquired modifications are common to both sexes, or at least to the individuals which produce the young." It is in this strong sense that Lamarck takes the maxim: *habits form a second nature*, though those who use the maxim do not see in it all that Lamarck holds to be there.[13]

Lamarck painlessly establishes without difficulty half of his proposition: the prolonged failure to use an organ entails its atrophy. The evidence for this proposition prevents him from seeing the lack of evidence for its positive counterpart: the need [*besoin*] to possess an organ ends by giving birth to it. One can follow the process of reasoning as far as the connection of the forms of organs with their habits,[14] but one loses his footing when one tries to comprehend the connection between the existence of organs and that of the needs which they satisfy.

Through an inevitable consequence, one, moreover, not expected, Lamarck's transformism ends in a debauch of finalism. Short of the substantialization of the needs in order to make them efficient causes, which Lamarck expressly refuses to do, it remains the case that the organs are born, grow, and form themselves *in order to* satisfy the needs of the organism. That an organ should strengthen itself by exercise is comprehensible, and, in any case, it is observable; but that an organ should be born simply because a living body has need of it is a quasi-magical operation. It is nevertheless by means of "observations" of this order that Lamarck claims "to demonstrate" that the continued use of an organ, and the efforts made to use it in novel circumstances, not only reinforce and enlarge it but even create "new ones [organs] to carry on functions that have become necessary."[15] How can we imagine the birth of a new organ as the effect of its exercise, since that which does not exist cannot be exercised?

Lamarck courageously defied the difficulty, and we owe to his speculative intrepidity two pages which Cuvier has often been reproached for having cited in his academic "Eloge" of Lamarck, but which no one can honestly claim Cuvier invented: the efforts which they make in trying to swim have extended the membranes which lie between the digits of ducks, geese, frogs, beavers, otters, and so on. On the contrary, the habit prized by certain birds of perching in trees has stretched out the digits and the nails of their feet in order to allow them to perch better. The most astonishing is the waterside bird [*l'oiseau "de rivage"* (sic)] which, "not liking to swim" and yet needing to go to the water in order to fish, develops stilt-like legs. Thus, "wishing to fish without getting its body wet," it keeps stretching its neck in order to get one which may be sufficiently long. Cuvier has not invented this.[16]

It cannot be said that Cuvier's critique is false, but perhaps he does not do justice to the intuition which formed the basis of the theories of Lamarck: that of the possibility of a universal transformism, a hypothesis before which we saw Buffon hesitate, then beat a retreat. Cuvier took for granted that this was an error: "It is clear that once these principles are admitted, the only other things that are required are time and circumstances in order for the monad or the polyp to end by gradually and indifferently transforming itself into the frog, the stork, or the elephant." No, not "indifferently," but let us move on. Cuvier reasonably adds: "But we understand also, and M. Lamarck does not fail to add, that there are no species in nature."[17] He is only wrong to harbor resentment toward Lamarck for this, for the very truth of this proposition was the principal object of his inquiry: if there had been no species created at the beginning by the author of nature, how could it come about that there appear to be species today?

In rereading Cuvier one notes the presence of two problems: the explanation, in fact imaginary, invented by Lamarck in order to lend reason to the formation of species, however provisional they may be; and the very fact that if one brushes aside as nonscientific the hypothesis of a divine creation of species, their existence requires a properly scientific explanation, which Lamarck's was not.[18]

From the theological point of view Lamarck occupied an irreproachable position. If God created the world, he created it such as it is. It is for science to say what the world is, and, whatever it be, the world of science is that which God created.[19] From the scientific point of view Lamarck proposed an explication, which is at least debatable, of the origin of organic variations, which are the origin of species. But perhaps the problem is metascientific? In fact, these organisms, endowed with the power of secreting modification of organs for which they have need, if not the organs themselves, have a strange resemblance to Aristotle's organisms, which, working from within by their substantial form, progressively shape their matter according to the type of perfected being which they tend to become.

How are we to explain "this admirable work of nature"? When it comes to first questions in this matter, Lamarck is content to speak everyday language, which is that of finality.[20] Nature *wanted*, nature *had to*, nature *had need:* these expressions and others like them are not infrequent in his writings. Still, it is necessary that nature predispose organisms, in order that they be conscious of providing beforehand for the organs of which they will be in need. This has been done, thanks to the primitive explosions of living matter, the reality of which Lamarck never doubted. One must admit that

> nature herself produces direct or so-called spontaneous generations by creating organization and life in bodies which did not previously possess them; that she must of necessity have this faculty in the case of the most imperfect animals and plants at the beginning of the animal and vegetable scales, and also perhaps of some of their branches; and that she only performs this strange phenomenon in tiny portions of matter, gelatinous in the case of animals and mucilaginous in the case of plants, transforming these portions of matter into cellular tissue, filling them with visible fluids which develop within them and setting up in them various movements, dissipations, restorations and alterations by means of the exciting cause provided by the environment.[21]

There is the difficult conjunction. How is it that this stimulating cause of the surroundings acts? On what does it act? The existence of a living matter would not be that of living *beings*. Even the marvel of spontaneous generation does not explain how that which it produces is organized, or how the action of the surroundings finds in the organic matter a latent desire to satisfy. We see, moreover, that being so honest, Lamarck does not claim to explain it. It is from this point on that he explains things. Hence the admirable zoological garden to which he introduces us and whose marvels he details with complaisance. This is the land of finality turned upside-down, but of finality all the same. Birds do not fly be-cause they have wings; they have wings in order to be able to fly as they desire to. The great principle of finalism is intact here: "The forms of the parts of animals and the usages of these parts are always perfectly in harmony," and nothing is more comprehensible, since it is the needs and the usages which have developed the parts.[22]

It is thus that the skin which unites the digits of the feet of water birds "contracts the habit of extending," that the legs of wading birds and their necks elongate, unless, like the swan whose legs do not elongate, they acquire the habit, "while moving about on the water of plunging their head as deeply as they can into it in order to get the aquatic larvae and various animals on which they feed."[23] Swans consequently do not make any efforts, such as wading birds do, in order to elongate their legs. It is in the same world of transformations that the anteater, "in order to satisfy his needs," sticks out his tongue so often that it acquires a considerable length, and even if the animal "requires to seize anything with this same organ, its tongue will then divide and become forked."[24] Perhaps what we wonder at most is that the same causes are capable of producing opposing effects if it is necessary to do so to obtain different results. "Nothing is more remarkable than the effects of habit in herbivorous mammals." Those who have much grass to browse on and consume large quantities of it daily become "elephants, rhinoceroses, oxen, buffaloes, horses, etc." Those of the herbivores who inhabit desert lands are "incessantly exposed to the attacks of carnivorous animals"; "necessity has in these cases forced them to exert themselves in swift running, and from this habit their body has become more slender and their legs much finer; instances are furnished by the antelopes, gazelles, etc."[25] Happy the ruminants for whom the abundance of grass at their disposal shelters them from carnivores!

Perhaps the philosopher can learn something from Lamarck, namely, that every adaptation can be interpreted as a finality, indeed even a double finality, according to whether one takes into consideration what adapts or that to which it adapts. "Because" is perhaps the inverse of "why," and vice versa.[26]

But why intervene between Lamarck and his reader? It is a meritorious action to say where he himself situates the novelty of his position. Before him it was admitted that nature, or its Author, had given to all animals bodily organizations which allow them to live under all the diverse circumstances where they would have to live. With Lamarck nature has successively produced the animals, from the simplest to the most complex, and to the extent that she has diffused them over the surface of the globe, "every species has derived from its environment the habits that we find in it and the structural modifications which observation shows us."[27] Let us not contest with him the paternity of this doctrine of which he is so proud; let us simply establish the fact that it has caused the finality of God's thought to descend into the interior of nature, where, moreover, even if one situates it initially in the mind of God, it would be rather necessary to conclude by rediscovering it.

2. Darwin without Evolution

Two names symbolize for the broadly educated public the problem of evolution: Lamarck and Darwin. It is generally known that these names stand for two ways of explaining evolution, but also for a basic agreement concerning the reality of the fact.

Nevertheless, one can read Lamarck without coming across the word "evolution." So far as Darwin is concerned, he wrote no book whose title proclaimed a study of evolution.[1] That does not prove anything, but it is as if the word "critique" did not figure in the title of any of Kant's works: It is curious. The word "evolution" does not appear moreover in the title of any of the fifteen chapters of the *Origin of Species* or of any of the twenty-one chapters of *La descente de l'homme.*[2] Darwin drew up brief summaries of each of the chapters to be printed immediately after their titles. In not one of the summaries of the thirty-six chapters does he speak of evolution. If one curious about history were to undertake the reading of the *Origin of Species* in order to find out what Darwin said there about evolution, he would ascertain with surprise that the word is to be met with nowhere, either in the first edition (1859) nor in any of the subsequent editions until the sixth, which appeared ten years after the first (1869), where the word finally appears in Darwin's hand concerning particular conditions which, because they present problems of design, cannot perhaps be completely elucidated.[3] We shall be tempted to say something about this, but the fact remains that Darwin himself did not have as his prime and principal purpose to promote a doctrine of evolution. He was able to present his work completely without using the word, of whose existence he was however aware. In a word, if there exists an inventor of the theory of evolution, it cannot be Darwin.

It is legitimate to object that what Darwin taught was the same thing that today we call evolution, yet it remains to be explained why, knowing the word, he so tardily and so sparingly made use of it. One could speak on that indefinitely, but the first response to make, one which explains at least in part the difficulty, is that at the time when Darwin elaborated his own doctrine of the origin of species, the word "evolution" was already in use to signify something completely different from what he himself had in mind.

According to its Latin origin, the force of which is felt in both English and French, "evolution," from the verb *evolvere*, would be the inverse movement of *in*-volution, the un-rolling of the in-rolled, the de-velopment of the en-veloped. In that sense, the only one which justifies the use of the term, it is an old philosophical notion, that of the *logoi spermatikoi* of the Stoics, grown into the *rationes seminales* of St. Augustine, St. Bonaventure, and Malebranche: In brief, the notion adopted by all those who wish to make absolutely certain that the divine act of creation having once taken place, nothing new is added to the created nature. St. Augustine loved to cite the text of Ecclesiasticus (18:11): *Creavit Deus omnia simul* [God created everything simultaneously]. Modern exegetes declare that this is a meaning contrary to what the text implies, but Augustine, who often exploited it freely, thought that even the contrary sense perpetrated in translating Scripture could at times be inspired. In any case, instead of understanding that God had created everything "without exception," Augustine and his school understood that everything that ever had been, or would be, had been created under a latent form, invisible, since the time of creation, which took place in the twinkling of an eye. Since everything that developed came from that, we have here a true doctrine of e-volution, understood in its natural sense of the un-rolling of something already given. It is in order to exclude the possible appearance of something new which should come into being without having been created that that doctrine of the *rationes seminales* had been conceived. This was a matter of a conservative creation. At any rate, the notion of a "creative evolution" is made by this contradictory and impossible.

The most representative of the advocates of this view whom Darwin knew was Charles Bonnet of Geneva (1720-1793), the author, among other writings, of *Palingénésie philosophique*, a work founded on the notion of the preformation of living beings in their germs. When he gave the title "Preformation and Evolution" to one of his chapters, Bonnet described the essentials of this doctrine: e-volution of the preformed, which is already given. Bonnet opposed a yet more ancient doctrine, that of Aristotle, which was taken up again in the seventeenth century by the admirable Harvey under the name of *epigenesis*. Neither Aristotle, nor Harvey, nor Bonnet posed the question of the origin of species or of their possible transformation. For Bonnet it was a question of what today is called ontogenesis, the development of the individual, in opposition to phylogenesis, or the development of the species. Thus one must choose between epigenesis, the doctrine according to which an organism grows from germs by the successive acquisition and formation of new parts (the position universally accepted today), and evolution, or the original preformation of future beings in the seed which needs only to develop. Bonnet explains himself in the matter following the title of his chapter: "If everything has been preformed from the beginning, if nothing is engendered, if what we inappropriately call generation is only the principle of a development which makes visible and palpable what was previously invisible and impalpable, then it must be the case that either the germs had been embodied originally one in another, or that they had been disseminated in all the parts of nature." Between the doctrine of the embodiment of germs and the kind of panspermism which he also considered, Bonnet does not choose firmly, but he inclines toward the first alternative. According to him "organic wholes have been originally preformed, and those of the same species have been enclosed one inside another. . . . The entire tree or animal, everything organic in general, is shown in miniature in a seed or an egg. A seed or an egg is, properly speaking, only the tree or the animal concentrated and folded up under certain envelopes."[4] This evolutionism of the individual, without any connection with Darwinism, was that which was discussed still in 1860 in the Academy of Sciences of the Institute of France. Its opponent, M. Serres, took Bonnet for a sort of fixist precisely because he taught an evolutionism warranting the unchangeableness of the individual, who is already completely present from the first instant of his evolution. In an amusing image Serres assimilates the doctrine of evolution according to Bonnet to the Old Testament of biology, and the doctrine of epigenesis to its New Testament.[5] Here it is antievolutionism which is the force for change.

In the circumstances which we shall attempt to elucidate, the sense of the word "evolution" had completely changed between Bonnet and Darwin. It had lost its first sense at least, the only one, to speak truly, which corresponds to it correctly, thus inaugurating an era of verbal confusion from which scientific language has not yet emerged. What certain contemporaries of Darwin called "evolution" was in fact its contrary, a sort of epigenesis, and as he himself taught a variation of it, it is conceivable that he may not have spontaneously formulated his theory of the origin of species in terms of evolution.

Nothing is less like Darwin's doctrine than the idea that new species should be already present in their ancestors, from which they only have to evolve in the course of time. Now, if the word "evolution" does not signify the contrary, or the inverse movement, from that of an in-volution, it does not signify anything intelligible. It is not certain that the present chaotic state of scientific evolutionism is not but the deferred effect of this original fault. Darwin at first avoided it. In a sense he was never personally responsible for it. The capital truth which he meant to show was twofold: first, that species have changed over time and, next, that they thus have been modified in virtue of a general phenomenon which he called "natural selection." These were his own doctrines, and this was also his language, so much so that he did not have to take into account that of others. He undertook right from the beginning to demonstrate that there had been a "transmutation of species,"[6] a term, let us note, much more appropriate to his thought than that of "evolution." Later on he will use the word "transformation" liberally, in the sense of a change of form, which would justify to a certain extent the epithet "transformist" attached to his doctrine. But his own manner of speaking is different. Rather than speaking of "transformism," he designates his point of view as that of "the theory of the modification (of species) through natural selection."[7] One could cite as many texts as one wants from him bearing out this sense. One has there his spontaneous thought and language, as reflective as can be expected from an impassioned observer of facts who was less interested in the choice of words. But, still, he had his own language. As soon as he begins to add to "natural selection" the words "or the survival of the fittest," one is no longer reading the first edition of the *Origin*. Spencer, we shall see, said the same thing; but this is Darwin's personal thought which he sets forth, without any adulteration.

Thus, in Darwin's own writings nothing announces the present complete fusion of sense between "Darwinism" and "evolutionism." Today they are identical.[8] One of Darwin's historians, however perspicacious, has noted, without musing on the astonishing fact, how infrequently the founder of evolutionism has spoken of evolution.[9] Assuredly, the question could be eliminated by admitting that there is no notable difference between the sense of the word "evolution" and that of the expressions used by Darwin. It is not necessary to enter into such a lexicographical discussion, for, as we said, Darwin was quite cognizant of the word, and it is necessary in any case to explain why, knowing it, he did not adopt it.

Let us recall briefly the known facts. The *Origin of Species* is the abridgment of an immense work, perhaps impossible to write, which Darwin for a long time hoped to bring to completion. He began to bring together his first notes on the subject in July 1837, after his return in 1836 from the cruise on the *Beagle*. From that moment on Darwin did not cease from working silently on his great work. He followed ideas so distinctly his own, so novel, and in his eyes so unbelievable, that it had never come to his mind that anyone else could have the same ideas. This is, however, what happened. At a time when Darwin, filled with scientific scruples, was amassing mountains of observations in favor of his own conclusion, an imaginative mind, endowed in addition with serious scientific competence, arrived independently and without such efforts at conclusions quite close to those of the naturalist of the *Beagle*. Darwin was thrown into confusion. In 1876 (?) (sic) the *Autobiography* will say: "This essay contained exactly the same theory as mine." His originality, he says further on, was in danger of being "destroyed." He even spoke of deciding not to publish his own book. On May 1, 1857, at a time when he received the memoir in which A. R. Wallace put together his own conclusions, Darwin wrote to the latter: "This summer will make the 20th year (!) since I opened my first note-book on the question how and in what way do species and varieties differ from each other. I am now preparing my work for publication, but I find the subject so very large, that though I have written many chapters, I do not suppose I shall go to press for two years."[10]

As always Darwin hesitates in the expression of his sentiments. In this same letter to Wallace he wrote: "I can plainly see that we have thought much alike and to a certain extent have come to similar conclusions."[11] Speaking of the article published by Wallace in 1855 in the *Annals of Natural History*, "On the Law Which Has Regulated the Introduction of New Species," Darwin says he subscribes to almost every word contained in the essay. It is known under what circumstances, well-advised by his friends Lyell and Hooker, he finally decided to publish jointly with Wallace two memoirs presented simultaneously to the Linnean Society under the common title "On the Tendency of Species to Form Varieties, and On the Perpetuation of Varieties and Species by Means of Natural Selection."

Let us note in passing that the word "evolution" does not figure in either of these two titles. In any case Darwin had been profoundly disturbed. He first was afraid of doing something dishonorable by claiming, as was true, that he had anticipated Wallace instead of Wallace having anticipated him. On June 29, 1858, he wrote to Hooker: "I daresay all is too late. I hardly care about it. . . . I send my sketch of 1844 solely that you may see by your own handwriting that you did read it. I really cannot bear to look at it. Do not waste much time. It is miserable in me to care at all about priority."[12]

He was troubled by it quite legitimately, however, but one wishes to know exactly why. Wallace's memoir was a recent composition which, we do not know exactly why, he had addressed to Darwin. Darwin's contribution was the essay of 1844 and an extract from a letter of September 5, 1857, in which, fortunately, he had explained his own theory to Asa Gray. The common title, difficult to arrive at, had at least the merit of allowing to appear, along with what the two points of view had in common, that which distinguished them: the natural tendency of species to form varieties was the common property of both authors, but the perpetuation of species by means of natural selection was the private property of Darwin. On exactly what grounds did he fear to lose his claim to priority?

The common response is the doctrine of evolution. But this is impossible, since neither he nor Wallace used the word in the memoirs or letters published conjointly under their names. It would be possible to be more precise and say "natural selection." But this is equally impossible, for natural selection figures in the title of Darwin's personal contribution – it is indeed the purpose of it – while Wallace's memoir does not mention it.[13]

A consideration of a totally different order ought to enter the line of account. Darwin's reaction is in large part that of a frustrated clergyman [*clergyman manqué*]. Probably for reasons of health, a certain measure of indolence characterized Darwin's temperament in general, as soon as it was not a question of observing plants and animals in their natural habitat. In order to find an honorable occupation for him, his father had thought to make a doctor of him. His evident deficiency of a medical vocation had next been interpreted by his father as meaning that he had a clerical vocation. The young Darwin loved to hunt, to fish, to botanize, to take long walks in the country collecting plants and insects, or even to observe geological formations of the land. None of all that appeared to him to be incompatible with the life of a village vicar. However, he wanted to assure himself before he took holy orders that he could in conscience subscribe to all of the articles of the creed of the Anglican church. Having assured himself that he could, he accepted the idea in principle, all the more so freely since, as he said in his *Autobiography*, "I liked the thought of being a country clergyman. Accordingly I read with great care *Pearson on the Creed*, and a few other books on divinity; and as I did not then in the least doubt the strict and literal truth of every word in the Bible, I soon persuaded myself that our creed must be fully accepted."[14] Let us note, for it is not unimportant, that he read attentively at the time Paley's *Evidences of Christianity* and his *Natural Theology*, which were given him, for their logic "was as pleasing as Euclid."[15] A long time later, on the *Beagle*, he amused his shipmates by citing the Bible in order to establish certain of his moral beliefs, but it was, however, his observations as a naturalist in the course of his long voyage which, dramatically opposed to what he held as to the literal truth of the Bible, destroyed his faith in the veracity of the Old Testament and, consequently, in all revelation.[16] Genesis claimed that God had created species by distinct acts of creation, and such have the species remained to this day. Since this was false, the Bible was not deserving of belief, and there was no more reason to believe anything for the sole reason that the Bible said so. From this moment on the religious beliefs of Darwin were progressively effaced. He never came to a declaration of atheism–absolute positions went against the grain of his nature–but he ended in agnosticism, which he retained right to the end, Westminster Abbey included.

The importance of this point cannot be exaggerated. Historians tend to overlook it because, after all, what Darwin thought of the Bible offers no scientific interest. But unless one takes account of it, it is difficult to explain his attitude toward the upholders of the doctrine of evolution. Even if they did not understand evolution, even if, as with Darwin and Wallace, they did not think it necessary to use the word, they were at least unified by a common conviction which made of them a sort of doctrinal party and conspirators against a common enemy. Some, such as Thomas H. Huxley, were pleased to think of it as such; others, such as Darwin himself, gave the matter much less thought. In fact, whether they wished it or not, they were all allies in the service of the cause of science against religion, of reason against faith in the revelation of Scripture. Darwin was at least aware of it. It was for him the occasion of a profound personal crisis, although it was not in his nature to draw romantic effects from it. He thought of himself as isolated in his spiritual struggle, at one and the same time troubled and proud to be the first to arrive at a conclusion of all the greater importance since, thanks to him, it would be henceforth scientifically demonstrated. His long hesitation to publish his conclusions perhaps depends in part on the importance of the religious truth which they bring into question. When Wallace proposed his memoir in which, for reasons other than Darwin's, but also scientific ones, he established the natural variability of species, Darwin felt his right of priority menaced on the point which, completely disquieting as it was, he held closest to his heart, and this made him decide to intervene.

Those who look into the matter with attention and according to its proper perspective will see that it is not a question here of an arbitrary historical interpretation. Darwin had been reproached, moreover, quite unjustly, with having been silent about his predecessors in the first edition of the *Origin of Species.* The reproach did not affect him, for concerning the strictly scientific part of his work, the theory of natural selection, he recognized few if any predecessors. But in the "Historical Sketch" prefixed by him to the third edition of his book (1861) it is precisely on the problem of scriptural exegesis that he is happy to find several predecessors.

> Until recently the great majority of naturalists believed that species were immutable productions, and had been separately created. This view has been ably maintained by many authors. Some few naturalists, on the other hand, have believed that species undergo modification, and that the existing forms of life are the descendants by true generation of pre-existing forms.[17]

It is necessary here to take care that the words "by true generation" are understood to mean "without divine intervention," for divine intervention would be a miracle and thus incompatible with the scientific spirit. Those who belong to the second category are the natural allies of Darwin. For example, Lamarck, for whose theory Darwin had at times very hard, almost injurious, words, but to whom he grants here an eulogy which deserves to be stressed:

> In these works (*Philosophie zoologique,* 1809; and *Histoire naturelle des animaux sans vertèbres, 1815*) he upholds the doctrine that all species, including man, are descended from other species. He first did the eminent service of arousing attention to the probability of all change in the organic, as well as in the inorganic, world being the result of law, and not of miraculous interposition.[18]

To eliminate all "miraculous" intervention here is to eliminate the creation that, in his imprecise theological terminology, he always held as a miracle, as if it were possible to have something miraculous in an act which, because it caused nature, preceded it. But this is of little importance here. Let us see rather what Darwin says of Spencer in this connection:

> Mr. Herbert Spencer, in an Essay (originally published in the *Leader*, March, 1852, and republished in his *Essays* in 1858) has criticized the theories of the Creation and the development of organic beings with remarkable skill and force.[19]

Neither there nor in the material following the notice does Darwin make allusion to the notion of evolution, regarding which he did not hold Spencer as a predecessor; not that Spencer had not spoken of it (he hardly spoke of anything else), but Darwin himself did not make use of it. On the other hand, Spencer criticized the doctrine of the creation of species by God. Darwin held him consequently as a predecessor and an ally, just as all the other anticreationists. Later, himself surprised by the rapid disappearance of creationist theory in his vicinity, he felt the need of convincing himself that it had been formerly as widespread as he had thought it to be. He never in the least doubted that he had shared the illusion. He would even hold it responsible for errors which he reproached himself with having made in biology when he was already in full possession of his principles. Two texts which bear out this sense merit being cited, one from the *Origin*, the other from the *Descent of Man:*

> As a record of a former state of things, I have retained in the foregoing paragraphs, and elsewhere, several sentences which imply that naturalists believe in the separate creation of each species; and I have been much censored for having thus expressed myself. But undoubtedly this was the general belief when the first edition of the present work appeared [1859]. I formerly spoke to very many naturalists on the subject of evolution, and never once met with any sympathetic agreement. It is probable that some did believe then in evolution, but they were either silent, or expressed themselves so ambiguously, that it was not easy to understand their meaning. Now things are wholly changed, and almost every naturalist admits the great principle of evolution. There are, however, some who still think that species have suddenly given birth, through quite unexplained means, to new

> and totally different forms: but as I have attempted to show, weighty evidence can be opposed to the admission of great and abrupt modifications. Under a scientific point of view, and as leading to further investigation, but little advantage is gained by believing that new forms are suddenly developed in an inexplicable manner from old and widely different forms, over the old belief in the creation of species from the dust of the earth.[20]

The change of tone is noticeable. We are now thirteen years and five revised editions beyond the first publication of the *Origin*. The *Descent of Man* has been published in the interval, and Darwin speaks freely now of evolution. He speaks of it as a "great principle," even though he had been able to write the *Origin* without mentioning it. He is indeed so convinced in the present that he believes he spoke of it twenty years earlier with a number of naturalists, even though the word does not appear one single time (to our knowledge) in his writings which date from that epoch. I am myself more opposed than anyone to the critical method which consists in believing oneself to be better informed than the authors one is studying concerning their real thoughts, but it must be admitted that the temptation to do so here is strong. If before 1859 Darwin had spoken so often of evolution with such a great number of natualists, how could it be that the word does not occur a single time in the editions of the *Origin* prior to the last one, the only edition to contain this passage? It appears that at that time Darwin admitted the existence of some great party of evolution, containing all those who rejected religious belief in a primitive creation of immutable, that is, fixed, species.

If he admitted the existence of such a party, Darwin could easily consider as already belonging to it, even if they did not yet use the word, all those who rejected creationism as the origin of natural species. From this moment he could depict himself to himself and others as having already discussed evolution, be it without naming it, each time that he spoke with others about the mutability of species. But it must be frankly admitted that this is a question of interpretation which the letter of the text alone does not justify. One is antecedently attracted to any better solution of the problem, with the sole reservation that that solution may not consist in saying that the problem does not exist.

The text taken from the *Descent of Man* is, at one and the same time, a perfect resume of Darwin's thought and a declaration of principle with which to interpret it.

> I may be permitted to say, as some excuse, that I had two distinct objects in view; firstly, to show that species had not been separately created, and secondly, that natural selection had been the chief agent of change, though largely aided by the inherited effects of habit, and slightly by the direct action of the surrounding conditions. I was not, however, able to annul the influence of my former belief, then almost universal, that each species had been purposely created; and this led to my tacit assumption that every detail of structure, excepting rudiments, was of some special, though unrecognized, service. Anyone with this assumption in his mind would naturally extend too far the action of natural selection, either during past or present times. Some of those who admit the principle of evolution, but reject natural selection, seem to forget, when criticizing my book, that I had the above two objects in view; hence, if I have erred in giving to natural selection great power, which I am very far from admitting, or in having exaggerated its power, which is in itself probable, I have at least, as I hope, done good service in aiding to overthrow the dogma of separate creations.[21]

A text such as this is inexhaustible. Let us focus on the pride which the former apprentice clergyman feels in having contributed, by publishing the *Origin*, to the ruination of the belief of creationism in biology, and thus to the consequent purification of science from this element which is foreign to its essence. It is because he had first to overthrow this obstacle in himself that he always attributed a considerable importance to the scientific decision which he had to take.

There was therefore in his thought a primacy of the problem of transformism over that of natural selection, which served only to give an explanation of the mechanism of transformation. The only alternative to the mutability of species (a scientific truth in his eyes) which he knew was the theological doctrine of creation. A letter of 1863 to Asa Gray allows no doubt in the matter:

> You speak of Lyell as a judge; now what I complain of is that he declines to be a judge. . . . I have sometimes almost wished that Lyell had pronounced against me. When I say "me," I only mean *change of species by descent.* That seems to me the turning point. Personally, of course, I care much about Natural Selection; but that seems to me utterly unimportant, compared to the question of Creation *or* Modification.[22]

The Bible or the transformation of species: such was, consequently, the basic option for Darwin from which he must proceed. This letter to Asa Gray is the only imaginable justification I know of concerning the claim of Francis Darwin in his edition of the *Autobiography* that as time went on, his father gave more importance to the recognition of evolution than to natural selection.

It is necessary to agree with this [claim] if one identifies the notion of the mutability of species with the notion of evolution, a notion to which most reputable naturalists have never assented. Charles Lyell, for example, of whom Darwin always spoke gratefully and deferentially, never accepted the idea that it was necessary to choose between fixism and the transformation of species. Cuvier did not admit it, either, but most remarkable is the fact that in this very letter to Asa Gray, Charles Darwin does not speak of evolution. It is Francis Darwin who thus translates the words written by his father in his most pure Darwinian language: *"Change of species by descent."* Moreover, it was in 1863 that he wrote these words, four years after the publication of the *Origin*, and not as the conclusion of a long reflection. He avoided the word, the sense of which appeared vague to him. He was merely in agreement with the anticreationism of those who used it.

But this text presents at least one other problem. Who are these partisans of evolution who at one and the same time reject the creation of species and natural selection? Many names might come to mind, perhaps that of Asa Gray, for example, who wrote in his critical recension of Darwin's work in 1860 that the doctrine of it was "largely accepted long before it was possible to prove it." This was to exhibit great perspicacity. Other names come to mind also, but the safest thing for us would be to turn to Herbert Spencer.

3. *Evolution without Darwin*

In a long citation from the *Origin of Species* we have allowed to pass an unusual expression of Darwin's: "The great principle of evolution." Darwin does not work on principles, except perhaps for natural selection, and it is certain that these words, which we come across late in the last edition, in the review of his last chapter, could not for him have emanated from the spirit of his first conception of the work. The passage in question[1] is, moreover, directed toward the past (1859), when almost no one believed in evolution, in order to set this state of affairs off against the present, when "things are wholly changed, and almost every naturalist admits the great principle of evolution." Why this novelty of vocabulary, and from whence comes this principle to Darwin?

One finds in the Bibliothèque of the Institute of France, under the name of Spencer, a brochure entitled "The Principle of Evolution," a response to Lord Salisbury, by Herbert Spencer. (An extract from the *Journal des Economistes*, number 15, December 1895. Paris: Librairie Guillaumin et cie, 1895). The title irresistibly suggests a connection with "the great principle of evolution" tardily inserted into Darwin's vocabulary [*la langue de Darwin*]. But one finds that this is not the English title of Spencer's essay, and, moreover, at this date Darwin had already disappeared from the scene. Since his death in 1882 he had entered into his glory.

In August 1894 the British Association for the Advancement of Science had held one of its regular assemblies. The president, Lord Salisbury, had taken advantage of the occasion to attack the modern doctrine of evolution, particularly under the form which it had taken in the philosophy of Spencer. The latter, as bellicose as Darwin was peaceable, drafted a response which he had translated into French and German, and distributed it in France and Germany as well as in England, "since there, as at home, it is necessary to make headway against reactionary ideas."[2] One perceives the novelty of the tone; we have decidedly gone beyond Lamarck and Darwin.

Nevertheless, Darwin will be involved, for, strange as it may be, Spencer's response, if not an attack on Darwin, dead now for twelve years, is at least an attempt to set himself off from his doctrine. Darwin counted for nothing in this affair. As for Spencer, he was simply the victim of the attack of Lord Salisbury, but the latter had mixed up the two causes, and Spencer could only disentangle them by emphasizing what distinguished them. It is not therefore the scientist who would distinguish his cause from Spencer's; it is the philosopher of evolution who wanted to distinguish his own variety from that, completely scientific, of natural selection. To know if the response of Darwin to the biological problem of the origin of species was true or not is a question the reply to which is beyond us. It is certain in any case that Darwin had posed a scientific problem, which he had long studied by scientific methods and to which, in his mind, the solution which he proposed has value only to the extent that it was scientific, that is to say, justified by reasoning based on the observation of facts. Darwin was the very incarnation of the scientific spirit, as avid in the observation of facts as he was scrupulous in their interpretation.[3] Hesitant by temperament, scheming when necessary, he fled publicity and detested controversy. Whatever his secret thoughts were concerning Spencer, and we shall know them before long, he was the last man to publicly implicate him, even if it were to separate himself from him.

Spencer was quite the opposite, but we shall see that he had excuses, which, moreover, had nothing to do with the person of Darwin.

One of his main grievances against Lord Salisbury is that he had confounded two distinct causes, Darwin's and his own. At the time of the incident, 1895 – that is to say, about thirty-five years after the first publication of Darwin's ideas – "Darwinism" already existed. That imponderable but invincible force, public opinion, already made of Darwin and Darwinism an event of planetary importance, at least within the limits of moderately enlightened opinion. We see Spencer himself, however irritated by the incident, speaking of "the coming of Darwin" as one speaks of one whose arrival marks the beginning of a new age, of a new era. Spencer resigned himself to the fact, but not without bringing forward several reservations.

In the first place he is astonished that so much importance should have been attached to the theory presented by Darwin. "Enthusiastic adherents have compared the principle of natural selection with the principle of gravitation."[4] The two cases are entirely different, and in order to show this difference, Spencer goes right to the heart of the problem: the difference in *nature* between his own absolutely universal theory of evolution and the particular, biological (indeed limited to a particular problem in biology) theory of Darwin's.

> Mr. Darwin's doctrine of natural selection and the doctrine of organic evolution are, by most people, unhesitatingly supposed to be one and the same thing. Yet between them there is a difference analogous to that between the theory of gravitation and the theory of the Solar System; and just as the theory of the Solar System, held up to the time of Newton, would have continued outstanding had Newton's generalization been disproved, so, were the theory of natural selection disproved, the theory of organic evolution would remain.[5]

The prime error of Lord Salisbury, that, at least, which directed the reaction of the philosopher, is, then, to have confounded two doctrines of different nature and reach. He confounded Newton with Copernicus. What is most remarkable is that in formulating this reproach, Darwin [sic; should be Spencer] recognized that at the time he wrote it, everyone already made the identification. Lord Salisbury "takes for his account the vulgar idea that Darwinism and evolution are synonymous terms."[6] He finally reasons as if the two notions were inseparable: "He assumes the two to be so indissolubly connected that, if natural selection goes, evolution must go with it – that the facts are not naturally explicable at all, but must be regarded as supernatural."[7] Without being aware of it, Spencer finds himself here revealing the profound accord which subsists between the two doctrines in his very decision to separate them. This is indeed what had led Darwin to accommodate the term "evolution," not to designate his own doctrine, but to signify his accord with those who, on whatever basis it might be, refused to allow the introduction into science of the religious, supernatural notion of creation.

Spencer certainly means to maintain his rights to this doctrine of evolution which the "vulgar" attribute wrongly to Darwin. That is Spencer's, and in order to establish his proprietary right, he reprints large extracts from an essay written by him "before the arrival of Darwin," at a time when "the hypothesis of development," as evolution was called, was universally held in ridicule. It should be noted in passing that the religious problem, or at least the theological one, is no less present to his mind than it is to Darwin's:

> In a debate upon the development hypothesis, lately narrated to me by a friend, one of the disputants was described as arguing that as, in all our experience, we know no such phenomenon as transmutation of species, it is unphilosophical to assume that transmutation of species ever takes place. Had I been present I think that, passing over his assertion, which is open to criticism, I should have replied that, as in all our experience we have never known a species *created*, it was, by his own showing, unphilosophical to assume that any species ever had been created.[8]

Spencer was so charmed by this restrained pleasantry that he cited it to console himself for not having had the occasion for making it. Furthermore, it recurred to him to say: if we do not have proof of evolution, you do not have any of the creation of species either. "Those who cavalierly reject the Theory of Evolution as not being adequately supported by facts, seem to forget that their own theory is supported by no facts at all."[9] This was true, but Buffon at least had not proposed the creation of the ass as a scientific theory.[10] What Spencer intended to emphasize is that before this time he himself rejected the position of the "believers in special creations,"[11] a theory so completely forgotten today that the historian risks not attributing to it a role as important as that which it actually played.

Whatever the case may be, it ought to be admitted that Spencer without doubt establishes the priority of his own theory (if not yet of evolution, at least of *development*) to that of natural selection. Darwin never posed as a champion of evolution. Spencer had no need to fear on this score.

Further on he stated that if in the two doctrines all particular creation of species is equally impossible, they are no less distinct for this:

> It is true that the contrast of evidences here emphasized refers not to the theory of the origin of species through natural selection, which at that time (1852) had not been propounded, but refers to the theory of organic evolution considered apart from any assigned causes, or rather as due to the general cause – adaptation to conditions. The contrast remains equally strong, however, if, instead of the general doctrine the special doctrine is in question; and the demand for facts in support of this special doctrine may similarly be met by the counter-demand for facts in support of the doctrine opposed to it.[12]

Everything is given at one and the same time in the texts, as it is in life. In defending the specificity of his own philosophical position, Spencer reveals incidentally his own scientific position in the matter of evolutionism properly speaking. Not only is it that Darwin did not teach evolution, but Spencer does not believe in natural selection. In claiming the paternity of the doctrine of evolution in general, and of organic evolution in particular, Spencer gives to it as its general cause "adaptation to conditions." In a word, even on the precise point of the cause and course of evolution, Spencer is not a Darwinian; he would rather be a Lamarckian. This separated him quite effectively from Darwin, for we know that the latter thought that Lamarckism was an absurdity.[13] The authentic Darwinian principle is not that of evolution; it is that of *the principle of selection.*[14]

Spencer's philosophy has today lost most of its credit. Its

claims lead one to smile, when one thinks that he, against whom Spencer asserted his rights, has become the eponymous hero of the nineteenth century, which has been called "Darwin's Century." But one must take Spencer's vantage point in order to understand it. From that time on public opinion was practically unanimous in speaking of the doctrine of evolution as Darwin's. It was perfectly reasonable for Spencer to protest and to reclaim for himself the paternity of the doctrine. But the confusion existed, and it was already hopelessly involved, for in large measure the great discovery which was popularly attributed to Darwin was not the evolutionism of Spencer, but his own doctrine of natural selection under the Spencerian name of evolution. Spencer had no less right to it from his own point of view, and he defined it in his memoir ["Lord Salisbury on Evolution"] in such precise terms as to leave nothing to be desired:

> How utterly different the popular conception of evolution is from evaluation as rightly conceived will now be manifest. The prevailing belief is doubly erroneous – contains an error within an error. The theory of natural selection is wrongly supposed to be identical with the theory of organic evolution; and the theory of organic evolution is wrongly supposed to be identical with the theory of evolution at large. In current thought the entire transformation is included in one part of it, and that part of it is included in one of its factors.[15]

In other words, it is popularly thought that evolution reduces itself to organic evolution, and that, in its turn, organic evolution reduces itself to natural selection, which, at best, is only one of the possible factors.

It will not then be useless to revisit now the little-frequented lines of descent of Spencerian evolution, and this time, it cannot be doubted, it is certainly a question of a doctrine of evolution, indeed, a philosophy of evolution, rather than a science of evolution, be it geological, as with Lyell, or biological, as with the neo-Darwinians.

Spencer is a philosopher, first, in that the goal he sets out is to obtain a totally unified knowledge;[16] next, in that he proceeds by conceptual constructions much more than by observation and description of facts. Armed with the idea of evolution, Spencer proceeds to the explication of inorganic, organic, animal, and human reality under all its aspects. He will not spend months in observing and describing a group of orchids or a colony of barnacles, as Darwin did. He has no need to go to the Galapagos Islands. It is not his style [*mètier*]. Truly a philosopher, Spencer starts from the universal in order to explain the particular.

All that is necessary to convince oneself of this is to open Spencer's *First Principles*. Starting out from evolution to speak of the phenomenon itself, he passes to organic evolution, which is a particular mode of it, and it is from thence that he will descend to properly human deeds which constitute philosophy, science, and art. Darwin and Spencer are like dog and cat; a sort of primitive discord separates them. Darwin cannot understand this abstract and verbal manner of speculating about nature, but one cannot have the least doubt but that Spencer's doctrine centers on evolution. In the notice in the "Historical Sketch" which we have already cited, after having praised Spencer for his vigorous critique of creationism, Darwin finds the means of praising the biological evolutionism of Spencer *without once using the word "evolution."* With a tact which one cannot but suspect of some malice, Darwin praises Spencer for having sustained on this point ideas which could be met with, if not everywhere, at least in many places. "He argues from the analogy of domestic productions, from the changes which the embryos of many species undergo, from the difficulty of distinguishing species and varieties, and from the principle of general gradation, that species have been modified."[17] After the brief allusion to Lamarckism which we have noticed above, Darwin mentions Spencer's work on psychology (1855), based on the principle of "the necessary acquirement of each mental power and capacity by gradation."[18] All goes on as if Spencer had never said a word about evolution, but those who are familiar with the cast of Darwin's mind know the reason for this silence. The "historical sketch" has as its precise purpose to render homage to the precursors of Darwin on the most important points of his own doctrine, and, in effect, there is not one of these points for which he praises Spencer that Darwin himself did not maintain in his turn. But he does not praise Spencer for having preceded him on the issue of evolution precisely because he himself, Charles Darwin, had not spoken of it. It is Spencer's doctrine, not his; he has then to recognize no priority in him on this issue.

The legitimate obstinacy of Darwin to hold himself to the strict terms of a sort of contract ratified with himself is a bit comical. His affair is natural selection; he has no need to speak about the rest. Having to speak about Spencer, however, Darwin did not find himself in a less paradoxical situation, since he must speak of

Spencer without evolution, while evolution is the very heart and head of the philosophy of Spencer. That philosophy is so little read today that perhaps it would be useful to recall the titles of several chapters of his *First Principles*: "Evolution and Dissolution," "Simple and Compound Evolutions," "The Law of Evolution," (two chapters), "The Law of Evolution" (concluded), and, finally, "The Interpretation of Evolution."[19] It would be difficult to argue that Darwin had failed to mention the evolutionism of Spencer through simple inadvertance. The most probable explanation is that he may have wanted to stay clear of this affair. Darwin does not mention Spencer as a predecessor in the matter of evolution because he himself is not yet engaged in the issue.

It suffices to turn to what Spencer said of evolution in order to see that the thought of the two men lacks a common measure. If we rejoin Spencer at the point where, after having defined evolution in general and distinguished simple evolution from complex evolution, he comes to organic evolution, the object of biology and zoology, we come across a definition which does not suffer from any ambiguity, at least so far as its intention is concerned:

> He will not forget that whatever aspect of it we are for the moment considering, Evolution is always to be regarded as an integration of Matter and dissipation of Motion, which may be, and usually is, accompanied by other transformations of Matter and Motion.[20]

One can imagine Darwin reading those lines, shaking his head, and asking himself: "How will that assist me in explaining the variations of form which I see in my barnacles?" Passages of this sort occur frequently; this one, for example, is from chapter 14, "The Law of Evolution": "Already we have recognized the fact that the evolution of an organism is primarily the formation of an aggregate, by the continued incorporation of matter previously spread through a wider space." In brief, it is a "concentration." Again, in a form more complete but of the same style: "[while] Evolution is always an integration of matter and dissipation of motion, it is in most cases much more." And, further on: "Whatever aspect of it we are for the moment considering, evolution is always to be regarded as an integration of matter and dissipation of motion."[21] For a biologist such as Darwin utterances of this sort were pointless.

Happily, we are no longer reduced to speculating about the personal sentiments of Darwin concerning Spencer, since Nora

Barlow restored a passage in the *Autobiography* of Charles Darwin which his son Francis Darwin had excised from the original without forewarning the reader. It is true that at the time when Francis published the autobiography of his father the situation had changed quite a bit. Public opinion had grown into the habit of glorifying Darwin, the illustrious author of the doctrine of evolution. Francis perhaps felt the awkwardness of publishing an unretouched judgment by the beneficiary of this popular error upon the true author of the doctrine of which he was popularly supposed to be the inventor:

> Herbert Spencer's conversation seemed to me very interesting, but I did not like him particularly, and did not feel that I could easily have become intimate with him. I think that he was extremely egotistical. After reading any of his books, I generally feel enthusiastic admiration for his transcendent talents, and have often wondered whether in the distant future he would rank with such great men as Descartes, Leibnitz, etc., about whom, however, I know very little. Nevertheless I am not conscious of having profited in my own work by Spencer's writings. His deductive manner of treating every subject is wholly opposed to my frame of mind. His conclusions never convince me: and over and over again I have said to myself, after reading one of his discussions,—"Here would be a fine subject for half-a-dozen years' work." His fundamental generalizations (which have been compared in importance by some persons with Newton's laws!)—which I daresay may be very valuable under a philosophical point of view, are of such a nature that they do not seem to me to be of any strictly scientific use. They partake more of the nature of definitions than of laws of nature. They do not aid one in predicting what will happen in any particular case. Anyhow they have not been of any use to me.[22]

This personal and direct testimony is unchallengeable, but one may understand why Francis Darwin suppressed it in the *Autobiography* if it is remembered that at that date, and for a long time before that, the international glory born of Darwin was that he had invented the Spencerian doctrine of evolution, or, at least, that he had invented a biological doctrine about which not many might have very precise ideas, but which, whatever it was, bore the Spencerian title of the doctrine of evolution.

Darwin was not much troubled by this misunderstanding. He was a modest man interested in nothing in this area except its research, its problems, and the solutions, always qualified [*nuancées*], which he thought it possible to propose concerning them. Spencer, on the contrary,[23] quite actively resented the situation. His doctrine of evolution triumphed under the name of Darwin, who had not taught it, with the paradoxical consequence that it was natural selection, which Spencer had not proposed [*voulait*], which usurped the title and the glory of evolution.

Even if it were no more here than a question of an interpretation of texts, that interpretation would be certain, more so in that the direct testimony of Spencer confirms it, and this testimony is more convincing in that Spencer himself, in giving it, prophesied that it would not change things. He spoke truly.

The "Preface" added by Spencer to the fourth edition of the *First Principles* is only a despondent protestation against the misappropriation of which he was the victim. Returning once more to his essays of 1852,[24] he reproached himself with not having said clearly enough for how long he held the theory of evolution, albeit under an abridged form. "No clear evidence to the contrary standing in the way, there has been very generally uttered and accepted the belief that this work, and the works following it, originated after, and resulted from, the special doctrine contained in Mr. Darwin's *Origin of Species*." Spencer next gives the dates and title of his first essays, which were later incorporated in *First Principles* and which, published before the *Origin*, could owe nothing to Darwin: "Progress: Its Law and Cause," published first in the *Westminister Review,* April 1857, corresponds to the materials in chapters, 15, 16, 17 and 20[25] of part 2 of *First Principles*; next, "The Ultimate Laws of Physiology," in the *National Review* for October 1857, without mentioning relevant passages in *The Principles of Psychology* (July 1855). Briefly: "As the first edition of *The Origin of Species* did not make its appearance till October, 1859, it is manifest that the theory set forth in this work and its successors, had an origin independent of, and prior to, that which is commonly assumed to have initiated it."[26]

On the surface Spencer does not have the least sense of the general difference which separates his philosophy and Darwin's science, but he at least sees that Darwin's science triumphs everywhere under the title of his own philosophy, and it is understandable that he does not find any pleasure in the fact. In any case it is too late, and that Spencer sees. "I do not make this explanation in the belief that the prevailing misapprehension will thereby soon be rectified; for I am conscious that, once having become current, misapprehensions of this kind long persist—all disproofs not withstanding. Nevertheless, I yield to the suggestion that unless I state the facts as they stand, I shall continue to countenance the wrong conviction now entertained, and cannot expect it to cease."[27]

This prophecy has come true. It is popularly asked who, Lamarck or Darwin, is the first inventor of the doctrine of evolution, although neither of them may have claimed the paternity of the discovery, while no one would dream of attributing it to Spencer, who claims it with good reason. This new unicorn,[28] *evolutionismus darwinianus,* gives proof of a remarkable vitality. It owes this, no doubt, to its peculiar nature as a hybrid of a philosophic doctrine and a scientific law. Having the generality of the one and the demonstrative certitude of the other, it is virtually indestructible.

What did Darwin himself think of it? That is difficult to say, for, unlike Spencer, this enemy of all controversy was not the man to put himself in opposition.[29] All the partisans of natural selection were partisans of evolution, in the sense that it was anticreationist. Taken together, they then formed one of the parties of thought which agree upon what they deny without necessarily agreeing upon what they affirm. Such is the case with many opposition parties. Darwin naturally found himself part of it, inasmuch as, in effect, one of his basic doctrinal positions, the most basic perhaps, had been the denial of the separate creation of distinct species. As this became a burning question concerning the origin of man, it is understandable that the word "evolution" should appear more often, or rather, less rarely, in the *Descent of Man* than in the *Origin of Species.* Then it is the anticreationism of Darwin which is expressed, a position which he admitted he held in common with others. He never claimed the paternity of the doctrine, but because it was a particular instance of the general problem of the origin of species, it should be allowed that he had furnished the scientific proof of it. Whether he did so or not is a question for science, and not for its history.

Francis Darwin did not have the same scruples in writing the biography of his father. Charles Darwin being then on his way to becoming the Newton of the nineteenth century for having discovered "the great law of evolution," the moment had been ill-chosen for an attempt to emphasize Spencer's rights of priority in the matter, and even for something more than simple priority, the invention of the law itself and of its name.

The honor of the family was at stake. No one has done more than Francis Darwin to consolidate the legend of Charles Darwin as the apostle of evolution.

> It will be shown, however, that after the publication of the *Origin*, when his views were being weighed in the balance of scientific opinion, it was to the acceptance of Evolution, not of Natural Selection, that he attached importance.[30]

This idea of a Darwin tardily converted from his own doctrine of natural selection and sexual selection to the doctrine of evolution appears extremely fragile. In the first place, except for an improperly interpreted passage of which we have spoken, one nowhere finds in Francis Darwin's book testimony signaling this important change of position on the part of his father. Next, and most important, the notion of such change makes no sense. No one, not even Charles Darwin, is capable of changing a strictly scientific view such as that of natural selection for a scientifically useless view such as that of evolution.[31] It is conceivable that in publishing the autobiography of his father, Francis Darwin eliminated the candid testimony to the low esteem in which Charles Darwin held the true inventor of evolution.

It is paradoxical that of two men so different in all respects, the modest one, always absent from scientific reunions devoted to the discussion of his work, should have gathered the glory for having taught a doctrine of which he himself, who well knew that it was not his own, hesitated to share the responsibility. It is Darwin, not Spencer, who had the national honors associated with a burial in Westminster Abbey. The Darwinism of evolution does not belong to actual history, but to mythological history. It is the fruit of a collective representation hereafter embodied in the press, in intellectual and political circles, and vivifying all sorts of vested interests.

One wastes his time today trying to correct the situation. One shall not succeed where Spencer himself failed. It is even probable that one will fail to have the real nature of the problem admitted. It is, one will be told, a matter of semantics. That was evolution which Darwin called *descent*. But precisely this is not true. Darwin never gave to his notion of "descent" the name "evolution." An excellent historian of Darwin writes that in the *Origin* the two theories "were supported by a single structure of facts and reasons, a structure so intricate that evolution could not be separated from natural selection."[32] In fact, what cannot be separated, either in the *Origin* or elsewhere in Darwin's works, is *descent* and *natural selection*. Natural selection is the cause of the descent of one species from another. It is, then, true that according to Darwin himself the theory of the origin of species would be unintelligible without that of natural selection, and since the origin is the first moment of the descent of species, it is indeed natural selection which is the kingpin of the entire work. Darwin completed his doctrine later on. He added sexual selection in explaining the descent of man. He even admitted that, though to a limited degree, adaptation to circumstances also contributed to the explanation of the descent of species. But he never substituted evolution for modification through natural selection. For him such would have been to renounce a scientific explanation and replace it by a word.

A final consideration may perhaps aid in perceiving the distance which separates the two doctrines. When one asks Spencer what he calls evolution, one obtains the verbal response which we have given: the movement of the homogeneous to the heterogeneous with dissipation of motion. This means nothing for the biologist Darwin. If one asks Spencer, again, the cause of the four or five classes of facts upon which is founded his belief in evolution (fossils, hierarchical classification, distribution in space, embryology, rudimentary organs), he responds that this cause is easily discovered. We only have to look around us to see everywhere in things a general cause which, if one presumes it to have been at work throughout all time, furnishes the explanation. Take any plant or any animal whatsoever, expose it to a new set of conditions – conditions not so different from the previous ones, however, that the change would be fatal – and (1) the plant or animal will proceed to change; (2) this change will be such that the plant or animal in question will finally be adapted to its new conditions.[33]

It is impossible to see anything more simple than this elementary Lamarckism. Between "descent," or the "transformation" of which the mechanism is natural selection, and the verbal explanation which Spencer calls evolution there is an entire lifetime of observations, comparisons, and classifications of facts connected by hypotheses – if not always propitious, then at least always reasonable and prudent. Darwinism and Spencerism do not speak to one another; they are worlds apart.

The fusion of the two doctrines under the title which has made it well-known is a social event well-designed to test the perspicacity of historians, but it is not certain that all the threads can ever be disentangled. The fusion was accomplished as early as 1878, in the remarkable article "Evolution" in the ninth edition of the *Encyclopaedia Britannica* (New York, 1878; vol. VIII, pp. 744-51). I do not dare to proclaim, but I am inclined to believe, that this article is in part responsible for the phenomenon which it describes, and perhaps that it even explains in part why evolutionism is a scientificophilosophical myth particularly lively in the United States of America. The synthesis in the article in effect divides evolutionism into two parts: "Evolution in Biology," the author of which is Thomas Henry Huxley, and "Evolution in Philosophy," entrusted to James Sully. We must, to our regret, limit ourselves to the contribution of Thomas Huxley, biologist, empassioned but competent and perspicacious witness of the event which he relates. The comments in his text are endless and tire the reader. We shall reprint only the main passage of Huxley's contribution, restricting ourselves to emphasizing solely those words which display the extraordinarily clever facility of this biologist when he dabbles in history. For him, let us recall, it is a question of making a place for Darwin in a history of evolution, of which he has so infrequently spoken. After having recounted the prehistory of the notion, Huxley continues:

> Nevertheless, the work had been done. The conception of evolution was henceforward irrepressible, and it incessantly reappears, in one shape or another, up to the year 1858, when Mr. Darwin and Mr. Wallace published their *Theory of Natural Selection* (of which Wallace does not say a word [Gilson]). The *Origin of Species* appeared in 1859; and it is within the knowledge of all those whose memories go back to that time, that, henceforward, the doctrine of evolution (of which the *Origin* did not speak [Gilson]) has assumed a position and acquired an importance which it never before possessed (in its new sense no biologist had yet spoken of it [Gilson]). In the *Origin of Species*, and in his other numerous and important contributions to the solution of the problem of biological evolution, Mr. Darwin *confines himself to the discussion* of the causes which have brought about the present condition of living matter, assuming such matter to have once come into existence. *On the other hand*, Mr. Spencer and Professor Haeckel *have dealt with the whole problem of evolution.*[34]

It is indeed necessary to introduce a philosopher among the scientists in order finally to find a theoretician of evolution! By a supreme act of ingeniousness, Huxley next compares Spencer to

Descartes in order to make us forget that Spencer had not invented analytic geometry and to give to his philosophical evolutionism a vague scientific tincture. We have here under our eyes the remarkable imbroglio from which issues the myth of Darwinian evolutionism. It is not born in the mind of Thomas Huxley alone; it appears to have sprung up a bit everywhere, as by a sort of spontaneous generation. But the article in the *Encyclopaedia Britannica* could serve it as birth certificate. It goes without saying that, in his turn, James Sully, the author of the section on evolution in philosophy, makes a place for Darwin among the philosophers. Filled with goodwill, he says that his theory, in sum, "is clearly a heavy blow to the teleological method"[35] (which we shall see denied by Darwin himself). But he, too, must eventually arrive at Herbert Spencer, "the thinker who has done more than anyone else to elaborate a consistent philosophy of evolution on a scientific basis."[36] And this time no correction or restriction is appropriate. Spencer is truly at home among the philosophers; evolutionism is truly a philosophical doctrine adorned with the plumage of science, but it is authentically a philosophy, and Spencer, not Darwin, is its author.

4. Darwin and Malthus

It is not difficult to discover the connections which at an early date tied together the thought of Darwin and that of Malthus. Darwin himself told the public of it,[1] but for a long time yet the sense and the bearing of Darwin's discovery of Malthus will be puzzled over.

The more one comes to know Darwin, the more one is persuaded that, from the day when he conceived the idea of transformation of species, he felt charged with the scientific mission of revealing to men a truth which was in his eyes indubitable; but this scientific truth was at the same time the reverse of a religious certitude which he himself had lost. The antireligious always has a bit of the religious in it. Strictly speaking, a scientific negation of the religious makes no sense, because the two orders are strangers to each other and because there is no sense of the word "truth" common to the two orders on which they might be able to meet. This abstract distinction is, however, contradicted by the psychology of the believer. There is in Darwin the scientist a propagandist charged by his own conscience with delivering men from a harmful error. Not having ever doubted the literal truth of the account of

Genesis, he was frightened, finding himself in the presence of his new idea. A world came apart, in his mind, under the pressure of its spirit. Many of those who today judge that his uneasiness was without objective basis would then without doubt have shared his fear. They are like those who in the twentieth century are astonished that it was possible in the seventeenth century to judge the theses of Richard Simon as dangerous to the faith.[2] At least Darwin had the courage to accept his own idea with all its consequences. In a letter to his friend Joseph Hooker dated January 11, 1844, that is to say, about fifteen years before the publication of the *Origin of Species*, Darwin said: "At last gleams of light have come, and I am almost convinced (quite contrary to the opinion I started with) that species are not (it is like confessing a murder) immutable."[3]

If species are not fixed, what is the cause of their variation? Darwin was much less able to neglect the question which had been posed before him by Lamarck, whose doctrine he knew well enough to feel authorized to reject as absurd. His own discovery of 1844 was not in his eyes that of the variability of species, for that uncovered to him simultaneously the cause of their variations. To depart from Lamarck had been to depart from a bit more audacious and technically perfected Buffon. Darwin himself only truly believed in the transformation of species when he was able to catch sight of the cause of their transformations, natural selection, which Lamarck had not imagined. The theory was virtually complete in his mind when he had discerned the essential parameters of the problem: the struggle for existence, the spontaneous variations in the heart of the species [*au sein des espèces*] with the tendency to divergence which they entail, the hereditary transmission of variations favorable to the perpetuation of the species, and finally the analogy between the results of natural selection and those of domestication.

The last characteristic is disconcerting, for to argue from domestication to natural selection is to compare a case of intentional and directed transformation to those cases where the cause of the operation is unknown. That stockbreeders put to profit certain spontaneous variations and favor them, to obtain a new variety, is a fact, and it is even a fact which is intelligible. A conscious process of selection goes on in stockbreeding, an intentional choice is made, *the end* of which is obtaining a new variety. It is the triumph of teleology. On the contrary, natural selection does not imply someone who selects. Darwin was fairly reproached for using the expression, though he thought the reproach unjustified. But he never completely renounced the use of the term, for it responded to a need of his intellect.[4]

Some have wished to define a purely scientific position on the problem by showing that the analogy of two selections, the natural and the artificial, is not an essential factor of it. In order to reason thus, the historian needs to substitute an ideal scientific problem instead of that which really presented itself to Darwin, and how can one be sure that one is not setting aside one of the necessary factors by doing this? In order to explain completely the formation of new species from spontaneous variations which have become hereditary, it is still necessary to explain orthogenesis, that is to say, to show why, or how, certain of these variations arrange themselves in a linear series, to result finally in new organs. Darwin did not wish either to content himself with chance or to invoke a single goal to explain this remarkable phenomenon, which is at the heart of the problem. He was disposed to speak about it only from a single analogical circumstance, that of domestication by horticulturalists and stockbreeders. Now these choose with intelligence; at times they select by a sort of genius. And to speak of natural selection is to speak of nothing if it is not to suggest that everything happens in nature *as if* one saw there the work of a selector, which one knows, however, is not the case. The notion is only extrascientific if one disregards the fact to which the notion corresponds.

We have seen Darwin assure us that he read Malthus *for amusement*, but this reading found him well-prepared to appreciate the doctrine of the struggle for existence. Already persuaded of the mutability of species, he sees immediately in the struggle for survival a means of explaining that it was possible for autoselection to proceed without a selector.

In the *Descent of Man* Darwin refers the reader to the memorable essay *On the Principle of Population, As It Affects the Future Improvement of Society* by the Rev. T. Malthus.[5] What of interest did he find there?

The first edition of the essay dates from 1798. Its author, the Reverend Malthus, belonged, then, to the clergy and presented himself as such. Himself an excellent man, without doubt even an upright Christian, he did not like the poor. It was not he who had written the celebrated sermon of Bossuet "On the Eminent Dignity of the Poor in the Church." Certain of his contemporaries were astonished at his sentiments: "Parson," William Cobbett addressed him contemptuously, "I have during my life detested many men, but never anyone as much as you."[6] He was not a detestable man; he was simply a man with a theory, that is, that the poor ought not exist, and if they exist, they do not have a *right* to assistance. Perhaps he made the mistake of expressing himself as if the poor themselves could do something about their poverty. His consolation was that in commiting them from their birth to parish nurseries part of the problem was resolved, since 99 percent of them died thereby in the course of their first year.

Malthus did not deny the fact, but this manner of doing away with the future poor appeared costly to him. The immediate cause of the evil was the Poor Law. The details of that law do not concern us; it suffices to know that the taxes imposed upon the non-poor for assistance to the poor had attained a level such that the contributors to it were exasperated. The parochial workhouses necessitated by the law were naturally in the charge of the clergy, and one would not be much deceived, perhaps, in thinking that the personal reaction of Malthus against the existence of the poor and the necessity of aiding them had not come to him despite the fact that he was a member of the clergy, but rather because he was such.

If the existence of the poor is prejudicial to the future well-being of society, what one does to come to their assistance, although doubtless humanly inevitable, ends by injuring the community. Malthus did not say that it was not necessary to sustain [*nourrir*] the poor; he only insisted that they had no right to be maintained, and, true or not, his proposition did not sound very evangelical.

The demonstration of the matter is very simple. It rests upon two postulates and one fact. The postulates are that (1) food is necessary to the existence of man and (2) the passion between the sexes is necessary and will remain nearly in its present state. The fact is that "the power which man has to populate the earth is indefinitely greater than that of the earth to produce sustenance for man." In meditating on this fact, Malthus proceeded even to propose a mathematical formula about it: "Population, when unchecked, increases in a geometrical ratio: subsistence increases only in arithmetic ratio."[7]

It is hard to say whether Malthus took his mathematical formula with complete seriousness; at least it was, to his mind, a striking manner of expressing the incontestable truth that, left to the natural play of conflicting forces, populations increase more rapidly than the means of their subsistence. At any rate he inferred from this that the Poor Law ought to be abolished because every law of this sort only perpetuates and multiplies the ill-adapted for whose existence it wishes to find a remedy. The measures taken by virtue of these laws work against nature, whose law is simply that people [*des gens*] for whom there is not sustenance do not have the right to exist. From that comes his conclusion, logically correct but not what one would expect from a man of the church and a Christian, that "we are bound by justice and honor to formally deny that the poor have a *right* to be succored." Assuredly, Malthus does not counsel the extermination of the poor, but he asks that an effort be made to secure from the poor themselves voluntary agreement to abstain from procreation.

This is to say that we live today in the age of Malthus. He would certainly be in favor of all contraceptive procedures, probably in favor of free, or even obligatory, abortion, in brief, in all legal measures for the limitation of births.[8] Living in a time which did not possess the means of limiting natural fecundity, he gave over such concerns to the methods of good advice, exhortation, and, if possible, persuasion—without deluding himself about the efficacy of these methods.

When marrying a couple belonging to the lower class, the clergyman ought to draw their attention solemnly to "the inconvenience, and even the immorality" of marrying each other without knowing whether they will be able to support their infants. If, despite this exhortation, a poor man does marry, as he has the right to, nature alone will be entrusted with punishing this fault, but the chastisement will be inevitable. The poor man who marries ought to foresee that he will have to suffer the consequences of his error. "He ought to know that the laws of nature, which are the laws of God, have condemned him to suffer, him and his family, for having defied their warnings; that he had no claim, no *right* upon society to receive from it the least parcel of sustenance above that which he is justly entitled to procure by his work, and that his only recourse is private beneficence, which does not go far." If parents abandon their infants, they ought to be held responsible for this crime. Anyway, the young "are, comparatively speaking, of little value to society, since others will take their places immediately."[9] Only the certitude which comes from the formulation of apodictic truth could give to Malthus the courage to set forth such principles so deliberately, as if the poor infant could be held responsible for the fault committed by those who have "inflicted life" on him.

But that was not what interested Darwin. He was particularly struck by that other Malthusian principle by which, in any case, nature herself necessarily eliminates most of what she produces. There are, in Malthus' first essay, passages which Darwin could not have failed to note. For example:

> Through the animal and vegetable kingdoms, nature has scattered the seeds of life abroad with the most profuse and liberal hand. She has been comparatively sparing in the room and the nourishment necessary to rear them. The germs of existence contained in this spot of earth, with ample food, and ample room to expand in, would fill millions of worlds in the course of a few thousand years. Necessity, that imperious all pervading law of nature, restrains them within the prescribed bounds. The race of plants and the race of animals shrink under this great restrictive law. And the race of man cannot, by any efforts of reason, escape from it. Among plants and animals its effects are waste of seed, sickness and premature death. Among mankind, misery and vice. The former, misery, is an absolutely necessary consequence of it. . . . This natural inequality of the two powers of population and of production in the earth, and that great law of our nature which must constantly keep their effects equal, form the great difficulty that to me appears insurmountable in the way to the perfectibility of society.[10]

One cannot read these lines without asking oneself why Darwin did not inscribe Malthus in the number of his predecessors in the "Historical Sketch" prefixed to the third edition of the *Origin*. It is probably that the problem presented by Malthus was not by nature biological and that, a moralist and economist, he had no place in a history of the origin of species. Malthus, whose problem was to know how to bring about the happiness of society while freeing the rich from the burden of sustaining the poor, did not set for himself any problem of selection concerning them. He did not search out in any of the poor signs of spontaneous variation deserving to be cultivated and transmitted by heredity. A Malthusian eugenics was possible on the face of it. Such did not take form, however, and it is really a kind of natural and spontaneous eugenics which Darwin described. Of all the readers of Malthus Darwin is almost[11] the only natualist who found in him that which he needed. If, as one is strongly tempted to believe, Malthus himself did not owe his observations to Charles Bonnet, there would be reason to see here a unique case at that date of a science of man serving as a pilot-science for a science of nature. But for Darwin Malthus did not count among the number of naturalists; he thus did not have the right to appear among the scientific precursors of his doctrine.[12]

There remains Malthus' passages where his doctrine of population is applied expressly to plants and animals. There it is applied with a necessity even more strict than to human populations. Man can struggle against overpopulation. Societies are capable of arranging things so as to produce more food, as they do moreover in our days. They can at least try to persuade individuals to reduce the number of conceptions and of births, a persuasion more especially efficacious as they put at the disposal of individuals more means of doing so. Nature, however, is incapable of preventing a vegetable or animal species from invading the whole earth–one would say today the whole universe. Nature contents itself with causing, by crude and haphazard means, a sort of autolimitation of species. She relies on the struggle for life to care for the permanent regulation of the multiplication of living beings by assuring the survival of the fittest and the corresponding elimination of the unfit. Darwin used a law of political economy for his own biological purposes. Even if he did once write the expression "struggle for existence,"[13] Malthus never thought of "natural selection," which remains the special property of Darwin.

5. *Evolution and Teleology*

Darwin had posed for himself the problem of the origin of species, but like others before him–Buffon and Lamarck for example–he had been led to think that species do not exist, that there are only varieties. Or at least there are moments when the zoologist, who observes and describes individuals as he sees them, holds a specimen as characteristic of a species which in the next instant he will hold as a case of a simple variety. Species themselves, in the first instance recognized as such, tend to dissolve into each other again afterwards, each departing from that which sets it off from others. "After describing a set of forms, as distinct species, tearing up my MS., and making them one species, tearing that up and making them separate, and then making them one again (which has happened to me), I have gnashed my teeth, cursed species, and asked what sin I had committed to be so punished."[1] The embarrassment was particularly inevitable since, as Darwin often reiterated, the continuity of the chain of being, each degree of which blends insensibly into those which preceed it and those which follow it, is one of the strongest arguments in favor of the transformation of species. As a critic will justly remark: "It is the origin of variations, whatever that may be, which is the true origin of species."[2] Now Darwin himself proclaims that variations are spontaneous, or even that there is in every species a "tendency to vary." One is not then quite sure that there are strictly definable species, and if one allows oneself to think that every so-called species is as it were a variety of another species, the problem of their origin loses all precise meaning. Insofar as species were supposed to be fixed, one could hope to know exactly what they were. There is no more place for a search for their origin from the moment that they cease to exist.

However, it is necessary to accept the Darwinian formulation of the problem if one wishes to understand the meaning of it. Let us grant then the spontaneous variations which stand at the origin of development. To explain the origin of species is not to explain the origin of these variations, since they are first, and hence inexplicable. What Darwin wants to know is how, these initial variations being given, living forms are constituted, endure, and even perpetuate themselves as one sees that they do.

Darwin, let us recall, always works on species already given. He asks himself not how it comes about that there are species,[3] but only how a species can give birth to another. But species, or varieties, are such complex structures that one can imagine only with difficulty that they could have existed at any time under a different form. If they occur, we scarcely understand how they could be produced. Darwin's attitude as regards this problem greatly resembles that of Aristotle: more discreetly than Lamarck, but so that one need not doubt the sense of his words, he leads his reader toward adaptations so marvelous that they amount to relations of final causality [*finalité*].

Not only without self-consciousness but with genuine delight, Darwin, like Aristotle, admires the beauty of nature. He is sensible, as we all are, of the brilliancy and diversity of colors in certain animals, above all in birds, and in a great number of flowers. But it is not this purely sensible beauty that is at question here. The often exquisite refinement of which nature gives evidence in the design of forms, the plotting of curves, and above all the unbelievable ability of adjustment in the adaptation of parts one to another, then the adaptations of the whole organism to the conditions of life in its environment, provoke in Darwin's mind a lively intellectual admiration for this intelligible beauty.

Concerning the first kind of natural beauty, and its appropriately aesthetic enjoyment, Darwin begins by denying that it has a theological origin and meaning. Certain naturalists "believe that many structures have been created for the sake of beauty, to delight man or the Creator (but his latter point is beyond the scope of scientific discussion), or for the sake of mere variety." If these doctrines were true, Darwin adds, this "would be absolutely fatal to my theory." With an inflexible steadfastness he refuses to dissociate the beautiful from the useful. Flowers, butterflies, birds, and numerous animals have become beautiful for the sake of their beauty, if one prefers, but "this has been effected through sexual selection, that is, by the more beautiful males having been continually preferred by the females, and not for the delight of man."[4] Sexual selection, which plays such an important role in the *Descent of Man*, then comes to be interchangeable with natural selection of which it is, moreover, if one might say so, only a variety, since the most beautiful male is also the one who has the greatest chance of perpetuating his line. Animal selection here becomes conscious, intentional, for even if the animal does not perceive it as such, it permits itself to be led by it as by a means toward an end.

The beauty which we wish to speak about is, however, rather of the second sort, that of the mutual adaptation of parts one to another and of the whole [being] to its environment. It is indeed a question of the intelligible beauty of observed interrelations, then, and Darwin is full of admiration in this regard. Amazement is the best description of his response: "You speak of adaptation being rarely visible, though present in plants. I have just recently been looking at the common Orchis, and I declare I think its adaptations in every part of the flower quite as beautiful and plain, or even more beautiful than in the woodpecker." Speaking of the *Orchis pyramidalis* and of the adaptation of its parts, he repeats: "I never saw anything so beautiful." And further on, still in connection with plants: "The beauty of the adaptation of parts seems to me unparalleled."[5] Darwin knows them better than Aristotle; his reasons for admiration are better founded, but it is the same admiration that is involved.

From this sentiment to the notion of final causality the distance is short. The beauty of adaptations is that of means to ends.[6] The adaptation of an organism to its surroundings and to its conditions of existence, and those of parts of an organism to other parts of it, are intelligible only from the point of view of their final result. That is what to be ad-apted means [*consiste*]. The transformist is not unaware of this, but he feels himself to be delivered from such widespread errors. The first and foremost of such errors is to conceive of natural finality as the result of an intention first present in the thought of God and capable, consequently, if one discerns it, of explaining the structure of his work. This theological finality is that of which Charles Darwin is the sworn enemy. No doubt is possible on this point. The second error, connected with the first, is to conceive of living beings as the result of any sort of fabrication. One deceives oneself completely when one reproaches Darwin with imagining natural selection as a choice brought about by nature. He desires, contrarily, a nature wherein all comes about as if there had been choice, even though no one and nothing were there to choose.

One comes then to the notion of a teleology without final causes. Solely through the play of natural forces, such as the tendency to vary spontaneously, the vital concurrence brought about by the scarcity of the means of subsistence and the elimination of the less apt which results from it, the ill-adapted forms eliminating themselves, the best adapted re-placing them, there takes place then a transformation of the former species and adaptation of the new ones to their conditions of existence more and more satisfying, without it being necessary to recur to the hypothesis of a causality of a particular type charged with directing the operation.

To speak the truth, it is hard to imagine the details of the operation. Spontaneous variations do not suffice to explain the initial structure, nor even the later modifications of structure. Actually observable plants and animals can only subsist thanks to the accord of the parts of which they are composed. What was that unknown ancestor ["X"] from which the woodpecker we know is descended? The latter could not subsist, or at least not easily subsist, if its tail, or its feet, or its beak, or its tongue were other than they are. How could its ancestors subsist without yet possessing the characteristics which presently assure the survival of the species? The evolutionists themselves recognize that, in the formation of a species, the first steps are the most difficult to explain, but that there should occur an entire series of concurrent spontaneous modifications to the constitution of a complex organism, that is an occurrence concerning the probability of which one cannot but wonder.

Darwin himself was not explicit on this point. He preferred to keep to himself such philosophical ideas as came to mind, but he was not opposed to his friends taking his place in order to put such ideas in evidence. In *Nature*, June 4, 1874, Asa Gray had published an article entitled "Charles Darwin," the reading of which gave Darwin immense pleasure, because it was elegiac and, above all, because it was intelligently so. Darwin did not have a great facility with the pen, and it could happen that a sympathetic and competent reader could give him the pleasure of finding his own thought more felicitously set forth by another than he himself knew how to do.

That was the case this time. Gray had said: "We recognize the great service rendered by Darwin to natural science by restoring teleology to it, so that instead of having morphology against teleology, we shall have henceforth morphology married to teleology." Darwin replied: "What you say about teleology pleases me especially, and I do not think anyone else has ever noticed the point. I have always said that you were the man to hit the nail on the head."[7]

It is curious that two men so intimately connected with Darwin's posthumous career, his son Francis Darwin, a naturalist like his father, and Thomas H. Huxley, he who made his reputation by being Darwin's "bulldog" [*bouledogue*], should have judged it useful to the glory of the master to emphasize the point in question. Thomas Huxley was the Darwinian "of the left," as eager for provocation as Darwin was reticent, and far from limiting the bearing of Darwinism to natural selection as also to other properly biological parts of the doctrine, he was one of those who contributed most to make of Darwin the champion of evolutionism. "For him, whoever he may be, who reads the signs of the times, the emergence of the philosophy of evolution, advancing with the mien of a claimant to the throne of the world of thought and freeing itself of detestable *impedimenta* which many would hope forgotten, is the most promising event of the nineteenth century." We see appear here, at least in outline, the mythical personage, today triumphant in the United States, who presides over "Darwin's Century." Now, it is this same radical evolutionist and avowed atheist whom Francis Darwin cites as having restored a sort of finalism. Here, to begin with, is the testimony of Francis Darwin himself:

> One of the greatest services rendered by my father to the study of Natural History is the revival of Teleology. The evolutionist studies the purpose or meaning of organs with the zeal of the older Teleologist, but with far wider and more coherent purpose. He has the invigorating knowledge that he is gaining not isolated conceptions of the economy of the present, but a coherent view of both past and present.[8]

In support of this personal opinion Francis Darwin cites these lines of Thomas Huxley, which have had hardly any attention paid to them:[9]

> Perhaps the most remarkable service to the philosophy of biology rendered by Mr. Darwin is the reconciliation of Teleology and Morphology, and the explanation of the facts of both, which his views offer. The teleology which supposes that the eye, such as we see it in man, or one of the higher vertebrata, was made with the precise structure it exhibits, for the purpose of enabling the animal which possesses it to see, has undoubtedly received its death-blow. Nevertheless, it is necessary to remember that there is a wider teleology which is not touched by the doctrine of Evolution, but is actually based upon the fundamental proposition of evolution.[10]

Those remarks are foreign to Darwin. He did not trouble himself with a philosophy of evolution; at any rate he did not feel responsible for such. It should be remarked, however, that two witnesses to his thought, including the watchdog who occasionally pestered him with his barking (as compromising as superfluous), had wished to note that evolutionism did not issue from Darwin, nor had his own doctrine of natural selection eliminated final causality. What exactly did he think of it?

It was a commonplace of traditional philosophy to put reason on its guard against the illusion of imagination. Kant was the first of those who denounced "the illusion of reason." It would be tempting to inscribe the judgment of final causality among the number of such illusions, but its case is different from those of the metaphysical ideas criticized by Kant. There is no sensible experience of the soul, the world, or God; there is, however, sensible experience of the facts which the understanding apprehends as connected by final causality, just as there is such experience of the facts which the understanding connects one to another by efficient causality. If one doubts that man, composed of his sensibility and understanding, perceives causalities by their appropriate acts, and therefore that he *sees* final causes as efficient causes, one can usefully reflect upon the personal experience of Charles Darwin.

He never completely exorcised the phantom of teleology. If it is an illusion, he was not able to deliver himself from it. In his remarkable letter of July 3 [1881] to W. Graham, the author of a book entitled *The Creed of Science*, Charles Darwin noted the great interest he had taken in the book, even though there were points in it with which he could not agree. "The chief one is that the existence of so-called natural laws implies purpose. I cannot see that." After having given his reasons, he makes this remark, which shows that he was quite cognizant of the issue: "But I have no practice in abstract reasoning, and I may be all astray." Then, after what has preceded, there comes an unexpected declaration: "nevertheless you have expressed my inward conviction, though far more vividly than I could have done, that the universe is not the result of chance." Coming to his senses, he continues: "But then with me the horrid doubt always arises whether the convictions of man's mind, which has been developed from the mind of the lower animals, are of any value or at all trustworthy. Would any one trust in the convictions of a monkey's mind, if there are any convictions in such a mind?"[11]

As a matter of fact, if there were such rational convictions in the mind of apes, they would be men, and their reasonings would be worth as much as ours. It is disconcerting that it is the author of the *Descent of Man* who has to make such a remark, for, in fact, according to him it is we who are the ultimate apes [*les singes finalistes*], and that if it be apes who come to this conclusion, it proves nothing against the doctrine of final causality.

An abstract from the remembrances of the Duke of Argyll reports what perhaps were the last words of Charles Darwin on this subject. They date from 1882, the last year of his life: "in the course of that conversation I said to Mr. Darwin, with reference to some of his own remarkable works on the *Fertilisation of Orchids*, and upon *The Earthworms*, and various other observations he made of the wonderful contrivances for certain purposes in nature—I said it was impossible to look at these without seeing that they were the effect and the expression of mind. I shall never forget Mr. Darwin's answer. He looked at me very hard and said, 'Well, that often comes over me with overwhelming force; but at other times,' and he shook his head vaguely, adding, 'it seems to go away.' "[12] Darwin naturally thought about that what everyone else does. He saw, as everyone sees, that the astonishing inventions which he himself had discovered in nature were, to return to the words of the Duke of Argyll, the effect and the expression of an elementary form of thought, or of a force related to thought, but because the evidence gave no purchase to demonstration, he turned his concerns away from it.

The long detour in which we have been involved with evolutionism will not have been useless. It allows us to see in the first place that the problem of final causality is just as unavoidable in the perspective of the evolution of species as in that of their creation. In fact, final causality today fares less ill than evolution.

The root of the difficulties is the fundamental indetermination of the notion of evolution. The notion signified something as long as it concerned the development of that which was supposedly enveloped, but Spencer popularized the word in another sense which no one could exactly define. Far from being the development of that enveloped, Spencer's evolution is a prodigious system of epigenesis where each moment adds something new to the one preceding it. One is already in creative evolution or at least innovative and progressive evolution. But whereas one understood an evolution in which the less issued from the greater wherein it was contained, that form of evolution in which the greater continually springs from the less is incomprehensible. It at least deserves no more to be entitled e-volution. One is not speaking anymore, then, of the evolution of a germ which already contained a tree, but of the rumbling of an avalanche which has nothing constructive in it. Words have their importance. *Evolution* has above all served the purpose of hiding the absence of an idea. The word was initially used to convey the meaning that all had already been accomplished in advance and then came to be used to say that everything that happened was new. In whatever manner biologists understand evolution, they are accounting for the mechanism of something the notion of which they are incapable of defining. It is necessary to be lost in the confusion oneself to get an idea of the logomachy of speculations called scientific devoted to the definition of this notion.[13] Only the decadence of scholasticism at the end of the fourteenth century offers a similar spectacle. It was inevitable that one came to doubt the reality of the very object of discussion. It is not so surprising that this should come about when one realizes that this last word of nineteenth-century positive science is the offspring of the crossing of political economy, a doubtful science, and the philosophy of Spencer, whose credentials are likewise doubtful.

Taking up the conclusions of the previous volume of the *Encyclopédie française,* devoted to living beings, the naturalist Paul Lemoine, professor at the Museum of Paris, stated that, instead of the definitive confirmation of evolutionism which was expected there, the contrary is to be found. "Volume IV of the *Encyclopédie française* will certainly mark an epoch in the history of our ideas on evolution. From its reading it becomes evident that this theory appears about to be abandoned."[14] The reason for this is simply that even those who accept it cannot explain either how it operates or in what it consists. "The theories of evolution with which our studious youth was lulled to sleep actually compose a dogma which everyone continues to teach; but, each in his specialty, zoologist or botanist, takes cognizance of the fact that any of the explications furnished cannot stand, whether it is a question of documents furnished by the Lamarckians, by the Darwinists, or by subsequent schools which appeal to these two great names." Natural selection, upon which Darwin relied to explain the changing of species, was nothing of the sort. On the contrary, "it has a conservative effect, and limits the variability of species, such that the individuals who are the most typical are those whom selection favors for survival." Paleontology, which was counted on to restore [*relancer*] the doctrine once again, was shown to be equally deceptive. Even in granting species 400,000,000 years in which to evolve, "the time is wanting for beings to evolve in, if indeed it is that they do evolve." In brief, and to allow the *Encyclopédie française* to conclude in its own words:

> The result of this exposé is that the theory of evolution is impossible. Basically, despite appearances, no one believes in it any more and one says – without attaching any other importance to it – "evolution" in order to signify "a series of events in time" [*enchaînement*]; or "more evolved," "less evolved," in the sense of "more perfected," "less perfected," because such is the language of convention, accepted and almost obligatory in the scientific world. Evolution is a sort of dogma which the priests do not believe in any more, but which they keep up for the sake of their flocks.
>
> It is necessary to have the courage to say that in order that

men of the next generation may direct their research in another way.[15]

It indeed would be agreeable for the philosopher could he take that negative statement as science's last word on the question. But the naturalists will not allow it to be so. Despite the firm conclusions of P. Lemoine, Jean Rostand believes one ought to maintain that "all the arguments given by Darwin almost a century ago remain perfectly valid."[16] That, however, is not all, for the same naturalist had clearly said, with less assurance: "It is patent that the great explanations of Lamarck and Darwin have in large measure failed."[17] One would like to know which among them remain valid. Of those of Lamarck, does one know any? But of those of Darwin even, how many can one hold as demonstrated?

It appears that it is a matter of no importance. Evolution has become so unquestionable that it henceforth takes the place of demonstration. Transformism presently occupies an impregnable position. "We are no longer in the times when, to make it acceptable, it is necessary to furnish a plausible explanation of the transformative process. It is the glory of the Lamarckian and Darwinian systems to have made the scientists believe in the idea of evolution. Necessary formerly to sustain a new-born transformism, such explanations can collapse today without damage to the process."

This is another way of saying that the theory, having passed from the state of an accepted prejudice, is henceforth, as is said, "in the air." In order to give some support to it, the same naturalist adds that, in any case, evolution is a fact: "as much as one can hold for a fact an event in which no one has assisted and which no one can reproduce."[18] But just as the indemonstrable is the contrary of science, so the inobservable is the contrary of the factual. Here one must excuse oneself and abandon the idea in order to follow the dialogue. The more one reads the scientists on this point, the more one is tempted to think that, like the notion of species, that of evolution is a philosophical notion which is introduced from outside of science, wherein it appears destined always to stand out as a foreign body.

CHAPTER IV

Bergsonism and Teleology

WHEN BERGSON WAS LED to take into consideration the notion of teleology, Paul Janet was hardly read by anyone, at least in scholarly circles. But Bergson himself had read him. His book on final causes belonged to a philosophy, still vaguely Cousinian, in any case prescientific, from which university teaching hoped to be definitively delivered.

The history of the book suggests nevertheless that the issue was not a dead one. Published in 1876, it was republished in 1882 in a revised edition, which was augmented with an important preface.[1] When, much later, we read the book in our turn, it was an agreeable surprise. Paul Janet was not at all the vague spiritualist we had been led to believe. Straightforward, sober, and lucid thought gave evidence of the desire to respect facts and not to confound biology and philosophy. Janet was acquainted with the objections to finalism, including those which were believed to have been drawn from Claude Bernard. He simply did not think the objections pertinent and proceeded to say why.

In 1882, in the "Preface" which he wrote to respond to his critics, Janet summed up his conclusions in the form of three propositions:

> First, there is no a priori principle involved in final causes. The final cause is an induction, a hypothesis whose probability depends on the number and the character of the phenomena observed.
>
> The second and fundamental proposition is that the final cause is demonstrable from the factual existence of incontrovertible combinations, such that the accord within these combinations, independently of the final cause, would be pure chance, and nature as a whole would be the result of an accident.

> Finally, the third proposition is that, after teleology has been accepted as a law of the universe, the only acceptable hypothesis which is capable of making sense of that law is that an intelligent cause is its origin.[2]

The third of these propositions is not of the same sort [*nature*] as the other two. We say, at least, that it goes beyond the limits of the present inquiry. In effect, it is a question of the proof of an intelligence which transcends nature. But even though this proof supposes the real existence of teleology in nature, the existence of natural teleology ought to be capable of being established on its own, independently of the eventual theological conditions of its possibility. Bergson did not devote any particular work to the problem of the final cause, but he could not fail to encounter it in *Creative Evolution*, and he discussed it there at length.

Following a traditional tactic, one at which he excelled, Bergson initially imagines two adversaries between whom he himself will define the proper position on the question. One is a *radical mechanist*, known since the time of Empedocles and a periodically recurrent phenomenon up to our own time; the other is a *radical finalist*, whom I have never encountered among actual biologists or philosophers.

Pure mechanism consists in maintaining that, extended matter and the laws of movement once given, the entire structure of the universe, including the living beings who inhabit it, and their history, can be exhaustively explained. La Mettrie would be a good example of one who held this doctrine, but Bergson cited the celebrated passage of Laplace, who gives a definition as complete as it is brief:

> An intellect which at a given instant knew all the forces with which nature is animated, and the respective situations of the beings that compose nature – supposing the said intellect were vast enough to subject these data to analysis – would embrace in the same formula the motions of the greatest bodies in the universe and those of the slightest atom.[3]

Bergson cites other analogous examples, one of which is from duBois-Reymond, and a third, which is most relevant because it is taken from Thomas H. Huxley, Darwin's "bulldog," and brings us to evolution:

> If the fundamental proposition of evolution is true, that the entire world, living and not living, is the result of the mutual inter-

> action, according to definite laws, of the forces possessed by the molecules of which the primitive nebulosity of the universe was composed, it is no less certain that the existing world lay, potentially, in the cosmic vapor, and that a sufficient intellect could, from a knowledge of the properties of the molecules of that vapor, have predicted, say the state of the fauna of Great Britain in 1869, with as much certainty as one can say what will happen to the vapor of the breath in a cold winter's day.[4]

These romantic professions of scientific faith make interesting reading in a time such as our own, where, provisionally perhaps, but as a matter of fact, the mind is not scandalized by the notion of a principle of indeterminacy which is difficult to reconcile with such prophetic determinisms as the above. They were not able to surprise Bergson, who knew Spencer too well to be taken unprepared by them.

Those who heard Bergson lecture on Spencer's *First Principles* at the Collège de France in one of his "little courses" perhaps took away the impression that it was all over with this kind of evolutionism. As Bergson himself would admit later on, "Spencerian evolutionism has to be recast just about completely," be it only to integrate therein the "real duration" which Spencer had excluded from it.[5]

However, there was something important which Bergson preserved from the evolutionism of Spencer, and that was the unreserved assent to the reality of evolution. Like Spencer, Bergson held it as a quasi-demonstrated certitude, and for the same reasons which we have seen alleged by Spencer:

> It is pointless to enter into the detail of observations which, since Lamarck and Darwin, have increasingly confirmed the idea of an evolution of species, I mean to say of the generation of one [species] by another from the most simple organized forms. We cannot refuse our adhesion to an hypothesis which is the beneficiary of the triple witness of comparative anatomy, embryology and paleontology.[6]

Present-day biologists would be happy to be able to share that triple certitude. Bergson had no doubt about it. He even thought that "science has furthermore shown by which operations all along the evolution of life the necessity of living beings adapting to conditions which are presented to them is translated." In fact, he inherited this notion of evolution neither from Lamarck nor Darwin, but from Spencer. Even if one should grant that Lamarck and Darwin were evolutionists without knowing it, it would be doing violence to their thought to attribute to them a notion, more philosophical than scientific, invented and popularized under that form by Spencer. Darwin was a biologist; in reflecting on the mass of facts which he had observed, he conceived the idea of explaining the transformation of species by natural selection; he was ready to do justice to other principles of explanation if the facts demanded it, but as a scientist he only thought within the bounds of the factual. When he dreams beyond such limits, he is conscious of doing so. Like Spencer, Bergson generalizes philosophically, in quest of a "completely unified" knowledge, on the faith of a science which he himself did not make and of which he has no personal experience. He borrows science from others.

It is permissible to ask if this fact has not affected his dialectical refutation of the evolutionisms of Lamarck and Darwin, such as he understood these to be. His purpose was to present a new and better philosophical interpretation of a doctrine which at the hands of Lamarck and Darwin insisted, furthermore, that it was primarily scientific. The "zoological philosophy" of Lamarck is nothing but the positivist manner of those whose most wild reflections remain in touch with facts which they themselves have observed. Bergson's admirable parallel critique of radical mechanism and radical finalism loses a great part of its pertinence when one realizes that it is a philosophical critique of scientific positions which their authors did not expressly universalize. Darwin thinks about spontaneous variation and of the effects of domestication. It appears to him that natural selection can explain them. In particular, he thinks that everything in this form of explanation makes more sense than does the theological belief in acts of separate creation. He simply says so, and that is all. More in the eighteenth century vein, Lamarck accommodates himself to an Author of Nature, and he enunciates his position more firmly than Darwin, but what is especially of interest to him is the fact that observable variations take place in plants and animals when their habitat changes. Bergson criticizes these two as if they were two possible moments in an evolutionist philosophy, which in fact only belongs to Spencer.

Spencer's presence often becomes obvious in formulas whose origin is beyond doubt. For example: "Yet evolutionist philosophy does not hesitate to extend to the things of life the same methods of explanation which have succeeded in the case of unorganized matter." Spencer said that [*l'a fait*], not Lamarck or Darwin. Checked in their inquiry, these mouthpieces of evolutionism conclude: "The absolute is not in our province; we are brought to a stand before the Unknowable."[7] Furthermore, it is Spencer who offers Bergson a more commodious target here than Lamarck and Darwin. We do not claim that the Bergsonian critique does not touch Lamarck and Darwin; it reaches them to the extent that their scientific thought is found incorporated by others in philosophical thought different from theirs in its method as well as in its goal. The evolutionism to which the Bergsonian critique is addressed, that of Spencer, excuses by its philosophical nature the "philosophism" of its refutation, but it is precisely an evolutionism as philosophical as Spencer's which Bergson proposes to establish in its place. In this sense Bergson is a continuation of Spencer.

This is what explains another characteristic of biological Bergsonism: like Spencer's, it is an optimistic evolutionism. Inspired without being aware of it by the optimism of Leibniz and Condorcet, Bergson confounds the two notions of evolution and progress. Optimism is not a necessary component of the idea of evolution. Even if it is appropriate to grant progress to the totality, it is very necessary to remember also that carnage is the rule in detail. Neither Buffon, who was rather aware of the "degeneration" of species, nor Darwin, who consoled himself with the thought that death most often comes quickly and with little pain, allowed themselves the luxury of a generous confidence in a bright and happy future. In contrast, the first writings of Spencer bear as their title *Essays on Progress*. Bergson will entitle his masterwork *Creative Evolution*. He never questioned his certainty that the universe was continually growing and that even the death of individuals might be "voluntary, or more or less accepted for the greater progress of life in general."[8] Between the "false evolutionism" of Spencer, which explains evolution by the products of evolution, and his own "true evolutionism," where reality is "followed in its generation and its growth,"[9] there are two common elements: both are philosophical systems, and both identify evolution and progress.

At the heart of this accord, the disaccord of Bergson with Spencer is vigorously affirmed. If there is evolution, it takes place in time. This is indeed why all biologists who speak of it are careful to assure themselves of enough time in which their scheme of evolution can take place. Now in Spencer's evolution time does nothing and counts for nothing. All the velocities could be multiplied by the same coefficient without the history of the universe being perceptibly modified thereby. Laplace's great formula would protect its complete verity in whatever time it might be, and if it were different from ours. This is a time without duration. Bergson often insisted on the fact that duration is the reality of time. Mathematical time is a translation of duration (Aristotle would say: of becoming) into the language of space. In terms of time thirty minutes are always equal to themselves; in terms of duration thirty minutes at a show differ from thirty minutes in the dentist's chair. The same time can appear to be more or less long. In boredom the condemned man says: "Time is hard on me" [*le temps me dure*]. It is on the evidence of this personal experience of duration that Bergson rejects radical mechanism:

> Radical mechanism implies a metaphysic in which the totality of the real is postulated complete in eternity, and in which the apparent duration of things expresses merely the infirmity of a mind that cannot know everything at once. But duration is something very different from this for our consciousness, that is to say, for that which is most indisputable in our experience. We perceive duration as a stream against which we cannot go. It is the foundation of our being, and, as we feel, the very substance of the world in which we live. It is of no use to hold up before our eyes the dazzling prospect of a universal mathematic; we cannot sacrifice experience to the requirements of a system. That is why we reject radical mechanism.[10]

It is at this point that, by a dialectic maneuver in which apparently he took great satisfaction, Bergson takes exception to radical finalism *for the same reason*. It is interesting to note that the type of radical teleology is, for him, the doctrine of another mathematician than Laplace, one who is at least of equal stature, namely, Leibniz. And it is quite true that in such mathematical finalism everything is determined in advance, everything is foreseen, and nothing new can be created. In that universe without creation or invention "time is useless again. As in the mechanist hypothesis, here again it is supposed that *all is given*. Finalism thus understood is only inverted mechanism."[11]

No reply can be found, and the success of the dialectical maneuver is complete, because an example has been chosen which fits the argument perfectly.

No middle ground having yet been conceived of between mechanism and finality, Bergson could only with particular difficulty maintain that one cannot condemn radical mechanism without throwing in one's lot with a finalism which, though not radical, is nevertheless a doctrine of teleology. No philosopher of nature worthy of the name pictured natural finality as producing living beings whose parts had been assembled according to a preconceived plan and in view of a certain end.[12] This time, further, Bergson is opposing an adversary who is not worthy of him, and it is tempting to give as a reason for this what Leon Brunschvicg said one day: "What weakness there is in Bergson lies in his choice of strawmen." He has allowed himself in this instance to be deceived by the example of an Aristotelianism he poorly understands, for one cannot believe that such a mind would voluntarily delude itself concerning this doctrine in order to overthrow it more easily.

Finality, he says then, "likens the labor of nature to that of the workman, who also proceeds by the assemblage of parts with a view to the realization of an idea or the imitation of a model."[13] To which he adds that mechanism proceeds in the same manner, which is possible, but the finalism of Aristotle does not proceed so. It is correct that, with Aristotle, the notion of final cause was inspired by the example of the artistic activity of the artisan or the worker, but it is not the case that mechanism lay the foundation for reproaching finalism with its anthropomorphic character.[14] In speaking about Aristotle we have insisted that it is art which imitates nature and not the other way around. What strikes Aristotle in comparing art and nature is precisely that, unlike art, nature does not calculate, reflect, or choose. This is indeed why, when nothing happens to disturb her activity, nature does not make mistakes. This is, finally, why, driven from within toward an end of which she is ignorant but which she carries about in her, nature does nothing in vain. Without models or trials, nature succeeds in the first attempt or she definitively fails. Nothing resembles less the work of the human artist guided by intelligence, for what characterizes that is the capability of being deceived. Nature works not "like the human worker, assembling parts," but by producing totalities whose existence implies the existence of what we call their parts. She does not build up plants or animals out of organs; she makes organs in the process of producing animals and plants. And she wills the parts in her willing of the whole. Like the God of Thomas Aquinas, nature does not will this in view of that, but she wills that this may be in view of that. It is significant that thought should feel the same need to escape to anthropomorphism in speaking of nature and in speaking of God.[15]

Aristotle often insisted on the fact that man works in the light of intentional ends with materials borrowed from nature, while nature herself produces her own materials. Man fabricates wings in order to be able to fly; he has not been capable of making wings sprout on him like those of birds, and this is moreover why, equipped with fabricated wings, he flies so poorly. Man has not discovered the secret of giving himself natural houses, akin to the dorsal and ventral shells of tortoises, but he has progressively learned how to construct them, and that is all that Aristotle said. *If* nature sprouted forth houses, her work would look like that of architects; but nature is not an architect, and her work does not resemble that of an architect. Her work is a natural being, and she herself is only an agent analogous to the intelligence which directs the operations of man toward ends which it conceives.[16]

The importance attributed by Aristotle to the fact that nature and art proceed equally by degrees, which implies the existence of an end, assuredly justifies in part the reproach that Bergson makes against him of entertaining an anthropomorphic notion of final causality. Nothing will show [*fera*] that Aristotle did not conceive one by analogy with the other. But here we should repeat on this occasion that man is a part of nature, that he is a unique case in nature, a nature which knows itself from within, and that through man who is part of nature she knows herself directly from within. Everything happens as if, in producing man endowed with reason, nature continued, under the form of production of the artisan, the work which she performed until then physiologically. It is a mistaken anthropomorphism to reason as if the two finalities worked in the same manner, as if nature fashioned an eye in the same manner that an optician fashions a telescope. But it is perhaps a legitimate anthropomorphism to think that two series of operations of analogous structure, and leading to comparable results, are in the last analysis of the same nature. Human craftsmanship continues the works of nature, and at times completes it, by entirely different means.

Perhaps Bergson himself was not, moreover, so far from the finalism of Aristotle as he imagined. Quite different from the finalism of the false Aristotelianism which he rightly criticized, Bergson's finalism is rather close to the truth. Evolutionism separates them. Aristotle, moreover, certainly never imagined the notion, hardly intelligible in itself, of a species which became another species. It would be better then, perhaps, to say, with Lyell, that one species disappears and that another is born from it; but how could it be proved that the first, in dying, engendered the second? Bergson speaks the language of evolution because it is the language of the science of his time: "the language of transformation forces itself now upon all philosophy, as the dogmatic affirmation of transformism forces itself upon science."[17]

We set to one side vitalism, which Bergson declares to be inseparable from the precedent position, although the biology of it is required today less than ever. It remains quite doubtful that the affirmation of transformism can be scientifically justified. All the species of animals known to Aristotle are still with us; not one of them has changed perceptibly in 2,500 years. If one thinks this period of time quite brief, one is naturally free to imagine what one will concerning the millions of years before then, but such belongs to the imagination. A contemporary biologist invites us to move further back in time without yet going to a fantastic distance. "It now appears with overwhelming evidence that the body and the brain of man have not undergone significant change in the course of the last 100,000 years." And he adds: "The same group of genes which governed the life of man when he was a paleolithic hunter or a neolithic farmer, still govern his anatomic development, his physiological needs, and his emotive drives."[18] Another biologist declares that "the parts of the brain phylogenetically ancient when compared to the neo-cortex have changed very little over the last fifty million years of the evolution of mammals."[19] Finally, commenting on his own testimony, the first of these two biologists remarks:

> All beings, having fundamentally the same structure, operate according to the same physiological processes and are moved [*poussés*] by the same biological needs. Nevertheless, there are no two identical human beings, and, a still more important consideration, the individuality of a person living today is different from that of all other persons having lived in the past or who will live in the future. Each person is unique, unprecedented, without a double.[20]

In this view evolution appears not to be oriented toward the production of new species each consisting of millions of individuals similar to each other, but through existing species to the production of innumerable individualities, irreducibly different. The *élan vital*, that creative push whose presence Bergson perceives in the origin of living species, appears then oriented otherwise than Darwin and the other biologists in whom he has confidence think. The struggle for life leads to species so stable that their crossbreeding with others becomes impossible and the proof of their existence rests in their sterility. In species individuals are so insistent upon the refusal to change that tissues from the organs of one, for a quite good reason, are rejected by those of another. Beyond a certain point of flexibility, beings forced to change simply prefer to die.

Bergson's great design was to put an end to the millenary conflict between mechanism and finalism. In fact, his own manner of conceiving the two condemned him, so to speak, to maintain a new version of finalism; his critique of Aristotle led him to revive true Aristotelianism and to restore to it the place usurped by false Aristotelianism.

It was not Bergson who invented inadequate finalism, wherein living beings only change in order to realize predetermined ends; but he ought perhaps to have made an effort to comprehend true finalism, that of forms immanent in nature and working from within to incarnate themselves there by modeling matter according to their law. His critique of intelligence, conceived of as originally flowing from the mold of action and, in order to prepare for the latter, completely occupied with proposing ends to itself, and then inventing the requisite mechanisms for attaining them, overlooks the possibility of an Aristotelian universe without Platonic ideas and without a Demiurge to impose them on matter from without. He then had reason to say that "radical finalism is very near radical mechanism on many points."[21]

In a time such as his, through an illusion which his own critique of rationalism assisted in dissipating, reason took itself to be intellect, and the finalism of the artist which he criticized could pass for the work of an essentially manual [*oeuvrière*] intelligence. This caricature of finalism deserved as a matter of fact to be criticized, but seeing that he rejected radical mechanism, he did not have any other choice than to have recourse to a certain notion of teleology purified of its vices. This new notion owed its novelty to what was a return to the ancient immanent teleology of Aristotle, less the forms which made the latter possible. This necessarily raised new difficulties for the doctrine.

We shall deliberately leave aside the problem of knowing what it is that Bergson incessantly calls "life." We shall suppose, in order to hold to what is essential in the matter for us, that he simply means by it the entirety of natural forces at work in living beings, and not a distinct energy such as those invoked by various vitalisms in order to explain specific matters in the vegetable and animal kingdoms. However that may be, Bergson speaks of it as of a *vis à tergo*, a sort of initial push [*poussée*], the *élan vital,* which, similar to a rocket [*fusée*], opens out in a shower of beings. The error of classical finalism would then be to have posited in advance the ends to be attained, whereas their principle and their eventual harmony were to be found henceforth retrospectively. A bit like the One of Plotinus, opening out into the intelligibility of *Nous*, the initial unity of the *élan vital* causes such harmony as there is in species: an imperfect harmony, certainly, but real, which constitutes a sort of consequent instead of antecedent finality. Bergson writes: "Such is the philosophy of life to which we are leading. It claims to transcend both mechanism and finalism; but, as we announced at the beginning, it is nearer the second doctrine than the first."[22]

Good philosopher that he was, Bergson had to utilize the notions that science put at his disposal. It was necessary for him here [to find] a sufficiently ambiguous notion which would allow him to navigate between two "radical" rocks, mechanistic radicalism and finalist radicalism. He found such a notion, naturally, in the notion of *adaptation*, which we have in fact ascertained sitting astride the two opposed doctrines. Bergson found in it the advantage of being able to explain thereby the existence of a harmonious whole [*ensemble*] without having to deny the discord which one comes across there.

It is a plausible philosophical notion as concerns organic evolution, but it appears that, preoccupied by his view of the global evolution of all nature, Bergson may have forgotten the more immediate problem of the formation of organisms. Intelligence, as he conceived it, was incapable of invention. It is a waste of time to recommence a discussion of the Bergsonian critique of intelligence; it is doubtlessly sufficient for our purposes to say that in order to be creative, evolution ought to be inventive, and that if it is necessary in order to conceive of it to compare it to its human analogue, it is of the creative imagination that we ought to think. A simple recipe for entering into contact with teleology in nature is to compose a sonnet. However poor the result may be, one will see there, in the work, the process in its totality, and, above all, in its reality.[23]

Far from having held teleology to be a peripheral notion, Bergson, then, sought to revive it under a purer form, and in a sense he succeeded, but not completely. With him the naive notion of the production of the present by the future ceased to exist. Even the more simple concept of an adaptation of the present, which exists, to the future, which does not yet exist, has been eliminated.[24] The energy necessary for the origin of all movement is, if not clarified, at least named. On the contrary, no attempt was made to resolve a more immediate problem: How is it that the *élan vital* explodes in divergent rays whose components are organisms?[25] Bergson could not in his response revive the notion of "substantial form." It was discredited and doubtless appeared to him to be certainly a return to the radical finalism which he wished to exorcise.

Such had not been the case however, since, rather than a model reduced to the dimensions of a future being, the substantial form is a plastic energy operating in matter in order to realize there concretely the idea which it is. It is assuredly necessary to loosen up Aristotelian fixism, but that ought to be possible since it is less a question of a position born of reflective choice than of a fixism, so to speak, by inadvertence. It is not forbidden to think of the [substantial] form as an inventive and at the same time conservative formula. Bergson, who so well understood the thought of Ravaisson, perhaps should have been able to find in a renovated peripateticism that with which to elaborate a solution to the problem that was at least acceptable.[26] But such considerations are basically of little importance, because being inevitable and useful at one and the same time, radical mechanism will always win the support of the scientific party. They will continue to pay lip service to it for a long time after they have ceased to believe in it. Because he left vacant the place formerly occupied by the substantial form, Bergson could do nothing efficacious to discourage the process.

It is he however who opened the way to a renovation of finalism. His remarkable failure to appreciate the true nature of intellect, which he obstinately continued to see as only the faculty of associating like with like, of perceiving and also of producing repetition–in sum, a calculating machine–leads him to situate elsewhere the source of invention, of creation, of all that by which the solution of a problem exceeds the simple sum of the items given. He consequently located this source in the vague entity he called "life," which he saw at work from the top to the bottom of the scale of living beings, up to man. In reflecting on it, he saw that there are some human activities, craftsmanlike in a sense, and therefore analogous to those which Aristotle cited as models of finality, but more exalted than the making of a couch, and for that very reason more capable of exemplifying a creativity similar to that of life. Artistic creation offered to his reflection the desired model. The free act offered a model no less satisfactory, but it is artistic acts whose structure and effects are more visible, more easy to observe.

It would be useless to ask Bergson to disown that which he held as the mainspring of his dialectic, that is, the unfitness of intelligence to create something new. Its natural calling is geometry. The mind can proceed in two opposite directions and consequently gives birth by its movement to two opposed categories. One of them, resulting from a sort of relaxation of the mind's natural tension, leads "to extension, to the necessary reciprocal determination of elements externalized each by relation to the others, in short, to geometrical mechanism." The other direction, which Bergson holds to be the "natural direction" of the mind, is, on the contrary, "progress in the form of tension, continuous creation." Having to situate teleology such as he conceives it, Bergson inevitably has to allocate it to the direction defined by intelligence, which is that of necessary determinism, repetition, and automatism. And what is to be said of the order of creative tension? In a curious phrase, which perhaps betrays a bit of embarrassment, Bergson says of this order that it "oscillates no doubt around teleology; and yet we cannot define it as finality, for it is sometimes above, sometimes below." It is notably in its highest forms, the free act or the work of art, that it is above teleology, for these manifest the perfect order characteristic of the relation of means to ends; and yet they can be analyzed in terms of means and ends only after the act has been completed or the work done. In a doctrine wherein teleology is only mechanism inverted, all that exceeds mechanism exceeds teleology.[27]

We ourselves say that teleology always exceeds mechanism, be it only in so much that it posits or implies the order to which it refers. Everything is mechanical in a machine, except the idea to construct it, which has dictated the plan of it. One hardly dares touch the luminous, translucent page wherein *Creative Evolution* develops perfectly self-assured views, nourished by truths of every sort, and nevertheless dominated by a kind of metaphysical Manichaeanism in which intelligence, dragging finality with it, is condemned to dwell in the house of geometry and evil. However, one would ask in vain of anyone other than Bergson for a perfect description of an intelligence creating the teleology and order which its work necessitates. Let us try, then, to go back up from extension to tension:

> Every human work in which there is invention, every voluntary act in which there is freedom, every moment of an organism that manifests spontaneity, brings something new into the world. True, these are only creations of form. How could they be anything else? We are not the vital current itself; we are this current already loaded with matter, that is, with congealed parts of its own substance which it carries along its course. In the composition of a work of genius, as in a simple free decision, we do, indeed, stretch the spring of our activity to the utmost and thus create what no mere assemblage of materials could have given (what assemblage of curves already known can ever be equivalent to the pencil-stroke of a great artist?), but there are, none the less elements here that pre-exist and survive their organization.[28]

What stands in the way of this analysis? Nothing, except the gratuitousness of attributing creation to "life" and the exclusion of intelligence which it presumes. Bergson is right, "we seize from within, we live at every instant, a creation of form," and this creation of form "is a simple act of the mind," which supposes in being at one and the same time the form, the matter, and the order of that matter which makes of it a poem. But this marvel works in us only because in us "life" is intelligence. There is life everywhere around us, and a poet could say that a tree is a poem, but he does not "write" trees. Bergson, who knew that moreover quite well, lost his way once on the way down from the Plotinian hypostases; he put "life" above intellect, the first-born son of the One. But if intelligence is, in us, the extreme advance point of life in the scale of known beings, it is by its position that it can conceive of life, and not vice versa.

The artists whose testimony Bergson invokes appear to be in agreement in thinking this way. Although their language is not that of philosophical technics, the most lucid among them direct our reflection in this direction.

Speaking of Joseph de Maistre, Charles du Bos wrote in one of his *Approximations*: "He possessed an exalted faculty, which the intellectual recognized and always deferred to, but which for others faded before the imagination in the current sense of the term: he possessed the ability to imagine ideas." In the course of the same essay du Bos mentions that "the ability to imagine ideas should not be confused with the scientific imagination; strictly, it is the imagination, not of the true, but of the intelligible."[29] And why should not the intellect have in effect its own power of setting up in itself, apart from abstraction, objects which transcend it? Principles are such objects, and they are forms.

Baudelaire was more of a poet than a philosopher, and, nevertheless, according to Gautier: "In his completely metaphysical conversation, Baudelaire spoke often of his ideas, very little of his feelings and never of his actions."[30] It is he who, speaking of the imagination, called it "the queen of the faculties,"[31] and one might think that which he meant to speak of was more than the imagination of images. There is no creative reason, but there is creative intellection. It is this intellection which is incarnated within language, whose forms it creates, including the forms of poems, which are verbal structures in which the poet creates at one and the same time the form, the matter, and the teleology which governs the structure. This work of creation is not necessarily conscious; rather, the testimony of poets invites us to think that, in large measure, it is not conscious. This is no reason to exclude it from intelligence. We do not know the extent of natural teleology. That which most closely resembles it is the creative power of intellect. It is not, then, absurd, it is even reasonable, to conceive of the cause of teleology as akin to intelligence. It is true that this is not a scientific proposition, but neither is its negation; and it would not be wise, out of respect for science, to deny such an important aspect of reality.

CHAPTER V

The Limits of Mechanism

WHILE FINALISM SURVIVED, mechanism came upon unexpected difficulties. In the game that has been played for twenty-five centuries between these adversaries, the stakes are not equal. Rare are those mechanists who admit that there may be teleology in nature, but exceedingly rare – if they have ever existed – are those finalists who deny mechanism and its natural function in natural beings. It could be shown that thus have things stood since Aristotle. He never denied that the mechanism of Empedocles was true, but he reproached him with presenting it as a total explanation of reality in the order of living beings, and contrary to Empedocles he insisted upon the presence of the "end" in living beings. Normally, mechanism excludes finalism, but finalism does not exclude mechanism. On the contrary, it necessarily implicates it.

It suffices to refer the reader back to Aristotle once more in order to convince him of this. According to him "there are two modes of causation, and . . . both of these must, so far as possible, be taken into account in explaining the works of nature, or . . . at any rate an attempt must be made to include them both; and . . . those who fail in this tell us in reality nothing about nature."[1] What Aristotle wishes to show is that the final cause, which is the primary cause of the whole operation, "constitutes the nature of an animal much more than does its matter." A couch, insofar as it is precisely a couch, is first of all an object contrived so that one could stretch out there so as to rest. Secondarily it is a thing of wood, of metal, or even of fabric and roping. This appears so evident to Aristotle that he does not succeed in persuading himself that the partisans of purely mechanical explanations have ever been able to blind themselves to this fact. "even Empedocles hits upon this; and following the guidance of fact, finds himself constrained to speak of the ratio [*o logos:* the French has *la raison*] as constituting the essence and real nature of things."[2] More evident still, it is of the essence of finalism to take into consideration not only the end of generation but further the matter and the mechanical forces arranged in accordance with the end.

It is a question here, not of a concession agreed to by finalism, but of a necessity. Volumes could be filled by citing testimony to this fact. Since it is necessary to choose, we shall consult him who was in the eighteenth century the universally respected representative of finalism, a theologian whose work we know came later to be familiar to Charles Darwin: William Paley.[3]

The first lines of the *Natural Theology* are indicative of what is to come, for one comes across there the instrument [*ce personnage*] which was destined to play a leading role in the modern history of final causality, the watch. We know what use Voltaire had already made of the clock in his satire *Les Cabales*:

> The universe troubles me, and much less can I think
> That this clock exists and should have no clockmaker.

Voltaire's clock perhaps engendered Paley's watch. If I stub my toe against a stone, and someone asks me how it comes about that the stone should be there, I would reply that I have no idea how it came to be there and that perhaps it has always been there. But if I stumble upon a watch, and someone asks me the same question, I would not be satisfied with the same reply. As a matter of fact, in examining the watch we see that, unlike the stone, "its different parts are framed and put together for a purpose, e.g., that they are so formed and adjusted as to produce motion, and that motion so regulated as to point out the hour of the day." Paley then proceeds to a detailed description of the parts of the watch and their arrangement in light of the foreseen end, to tell the time of day at any moment when one has need of knowing it. It is, then, the observation of this mechanism (*this mechanism being observed*) [Gilson] which alone allows one to infer that an artisan exists "who comprehended its construction, and designed its use."

Among the remarks with which Paley accompanies his argument we shall cite: the fifth, which says that it is not sufficient to invoke a "principle of order" in order to explain the watch, for a principle presupposes an intelligence to conceive it; the sixth, "that he [who stumbled upon the watch] would be surprised to hear that the mechanism of the watch was no proof of contrivance, only a motive to induce the mind to think so"; seventh, he who finds the watch would be surprised to learn that it is only the result of the laws of metallic *nature*, for a law produces nothing without a cause to bring it into play. The expression "law of metallic nature" may appear bizarre, but, Paley observes, it is the same as speaking of "the law of vegetable nature," "the law of animal nature," or simply of the "laws of nature" in general, as if laws were capable of causing that which comes about without an agent to put them to work. Our man would certainly not be put off his mark by the objection that he knows nothing of the matter. "He knows enough for his argument: he knows the utility of the end: he knows the subserviency and adaptation of the means to the end. These points being known, his ignorance of other points, his doubts concerning other points, affect not the certainty of his reasoning. The consciousness of knowing little, need not beget a distrust of that which he does know."[4] We only call on Paley's testimony and example here in order to confirm a rule which is, in other circumstances, self-evident: it is the simple presence of a mechanism which requires that we have recourse to final cause. This is why Charles Darwin will so often use the arguments and examples of Paley in confirming his own conclusions.[5]

Thus it is that, contrary to what we most often imagine, the substance of finalist reasoning is exactly the same as that of mechanist reasoning. The most attentive mechanists recognize the fact after their fashion, which is, not to deny teleology, but to try to give it mechanist explanations, taking the risk of falling back in the last resort on chance as an explanation of the living organism, although chance is the refusal to give an explanation rather than an explanation. It is not superfluous to examine this old doctrine of chance, already rejected by Aristotle, under one of its modern forms. How should we know otherwise whether it has not become true in the meantime?

The principal scientific event to take place in the twentieth century, at least up until the present, is, along with the theory of relativity, the physics of *quanta*. According to that doctrine energy is neither radiated nor absorbed in a continuous fashion, but in the form of discontinuous units called *quanta* of energy. A new microphysics was born which cast a new light on the elementary phenomena of life. As much as we can judge of the matter after occasionally confused controversies among philosophizing scientists, physical causality and its determinism remain intact, but there appears to be in certain domains of modern physics a sort of determinism which cannot be foreseen. Extremely low on the scale, where physical phenomena are produced, laws become, in some fashion, statistics. They explain the means and allow of a coefficient of indeterminacy, small but real. To know whether the indeterminacy is in things themselves, or only in our ways of observing them, is a point of great importance, but it is up to the physicist, and not to the philosopher, to decide what the sense of his science is. We can allow him then the burden of saying, if there is indeterminacy, on what exactly it has bearing. It does not appear, moreover, that the decision, whatever it may be, ought to affect the course of our own considerations.

We shall take as a guide one of our contemporaries, the American biologist Walter M. Elsasser, professor of geology and biology at Princeton University, author of *Atom and Organism: A New Approach to Theoretical Biology* (Princeton University Press, 1966). Educated in theoretical physics, known for his contributions to geophysics, he has also posed the curious problem of knowing what modern physics can say to us on the subject of biology.

In order to save his reputation, we hasten to say that Professor Elsasser is not a metaphysician. The only time that he uses the term "metaphysics" I distinctly fear that he uses it in the sense of "unreal." What holds his attention, moreover, is not directly "teleology": so far as I can determine, this word does not appear once in the book. "Vitalism" is what holds his attention, and this time it is my turn to feel not concerned.[6] The notion of "life" is Platonist, not Aristotelian. Assuredly, Aristotle often speaks of *zoe* and of the operations of life, but it is for him simply the proper action of living beings, that is to say, of beings which have in themselves the principle of their own movement. He never intends by this word a distinct principle, a force, an energy to which science or philosophy ought to have recourse, as to a cause, in order to make reasonable what we call the facts of biology. The problem raised by Walter Elsasser concerning these facts is of the greatest importance for us, because in discussing vitalism he is led to oppose it to mechanism and to say what he thinks of the latter.

Here is the principal proposition of his book: "The time-honored dualism of the mutally exclusive systems of thought, mechanistic biology on the one hand and vitalism on the other, express a pair of theoretical approaches which are both inadequate. We shall show how they can be replaced by an abstract descriptive system of a different type that is far better adapted to the nature of biology."[7] It goes without saying that we do not claim to take a position in that debate. The author states precisely, moreover, that his personal attitude in regard to the problem is that which "the modern scientist designates as positivistic." So much the better; it is not possible to be too positivistic in scientific matters. If, however, I should risk playing the prophet, I would warn our scientist [Elsasser] that he would have a hard time convincing his confreres in biology that he is a scientist rather than a philosopher, at least on the point in question. They will never hold as an authentic interpreter of scientific positivism a colleague who feels himself obliged "to move far away indeed from conventional mechanistic thinking."[8]

Nothing can take the place of reading a book such as this, but it ought to be possible to give an idea of its general tendency.

It is necessary first of all to decide to hold with the author that the theory of *quanta* is the last word, at least provisionally, of contemporary physics. Thanks to that theory, there is no further obligation to choose between two contradictory theories of light, that of waves and that of corpuscles. "Quantum mechanics has taught us that these two theories can be construed as two different and no longer contradictory aspects of reality; the dominance of either of these aspects is relative and depends on the method of observation." Then he adds: "Niels Bohr pointed out, first in 1933, that physicists discovered here a conceptual scheme of remarkable breadth and capable no doubt of further generalization, especially in biology."[9] We shall retain from these remarks only the fact, important for us, that Elsasser intends to proceed according to the method of physics, and even of mechanism, since the mechanism of *quanta* will be put in play; but he also proposes to show that biology ought to follow other ways than those of traditional mechanism.[10]

What stands in the way? Simply that it does not accord with the facts. There are laws of biology which do not allow of being deduced from those of physics. This proposition, which he himself says constitutes the center of his inquiry,[11] is immediately disconcerting. It is hard to imagine how it could be true, for it implies that one ought to expect to come across some general laws of biology endowed with *"a logical structure quite different from what we are accustomed to* in physical science."[12] Now this is precisely what is difficult to imagine: how could it be that natural laws founded on the known laws of physics and chemistry would present a "logical structure" completely different from that of these same laws?

His point of departure being physics, our biologist emphasizes initially a fact too often overlooked, although quite important: in an atomistic type of explanation all the atoms and all the molecules of a given kind ought to be exactly *alike*. Quantum mechanics is still more exacting. It shows that "without the indistinguishability or *quantitative identity of all electrons,* chemical bonding as we know it would not be possible." And further: "It is of course not at all obvious that Nature should be thus constituted. We could conceive in principle of forms of matter where any two atoms are a little bit different from each other. If these differences were large enough the standard procedures of physics and chemistry would be very much altered. It would no longer be possible to look up, say, the melting point or the absorption spectrum of a chemically pure substance . . . the structure of matter would look rather different depending on whether or not we are entitled to postulate the basic equality of its constituents."[13] Departing from there, our biologist proceeds to a series of surprising declarations, which even a philosopher hesitates to receive as "scientific" they are so generalized. This is all the more reason for relating them with an exact fidelity.

First: "Radical inhomogeneity is by universal consent an outstanding and altogether basic property of all the phenomena of life."[14] To support this statement W. M. Elsasser recalls the popular saying "no two blades of grass are ever alike." Leibniz loved to make the same remark about the leaves of trees. Without raising the least objection, let me make the observation that it is not perhaps a good example of scientific truths, for the proposition is unverified and unverifiable; but, to be precise, what our physicist wishes to say is that since the advent of the quantum theory of elementary particles, this traditional commonsense view has taken on a precise scientific sense. Modern physics does not consider anymore individual particles, or atoms, but rather classes, precisely because classes can be considered as homogeneous from the point of view of statistical physics, although their elements may not be homogeneous.

A second striking proposition of the same physicist (who modestly attributes it to Pascal) is: "Organic life . . . is inserted into inorganic nature in such a way that the former is of negligibly small extent compared to the latter."[15] And from this there naturally comes another question: How can the existence of organic nature be explained by "the existence of strict mechanical causality of the Newtonian pattern," or even simply be reconciled with it?[16]

Niels Bohr had already posed the problem, but Elsasser knows that its history goes back further still. Among earlier views proposed by biologists surprised by the exceptional nature of life, he cites as a remarkable example that of Claude Bernard. Elsasser recalls with perfect accuracy that, according to Claude Bernard, "there can be no deviation at all from the laws of physics and chemistry in the organism. He tells us again and again that physics and chemistry must ultimately be able to explain every detail in the functioning of the organism, but at the same time they cannot explain its existence."[17] Is this a likely situation? Is the mind capable of contenting itself with a view of living nature wherein the rules which explain its functioning fail to explain its existence?

Returning to this point [*Rendu en ce point*], our scientist proceeds, in the least expected manner, to a sort of profession of faith, or, to speak more simply, to a taking into consideration of commonsense, that source of information which science taught us to hold in suspicion in the future and even to contradict. Who does not remember the movement of the earth, the antipodes, and other analogous cases? Nevertheless, before this divergence between, shall we say, the essence of the living world and its existence, our scientist hazards a remark which it is fitting to cite in its entirety in order to be sure that we do not misrepresent it:

> [In the second place] the problem lies at the confluence of several special sciences. Specialists, by their very nature, tend to be selective. Philosophers on the other hand have always considered it their particular business to counteract the tendency of mental selection to which the practitioners of concrete sciences are often more exposed than the public realizes. If, therefore, all deal with the relationship of organic and inorganic matters we should relate ourselves to some degree to philosophy and to the continuity of the philosophical thought of the past. We could suitably apply an old maxim of the philosophers: after everything is said and done in philosophical analysis, the end result should not differ too violently from the short-cut solution offered by common sense, otherwise it might be philosophy rather than common sense which is suspect. Similarly here; we would do well to remain in touch to some degree with traditional philosophy. If the outcome of our inquiry is too much in contradiction with the somewhat intuitive results of traditional philosophy, it might not ultimately be tradition which is wrong.[18]

For a philosopher accustomed to reading scientists this is like a breath of fresh air. But he reads it with some amusement, in particular Elsasser's conclusion: "Bernard's view that neither vitalism nor mechanism can win a complete victory in their long drawn-out struggle seems only rational in this light."[19] The conclusion in question is not that to an interminable conflict, nor is it further dictated by the desire not to lose contact with the intuition of commonsense. We have already come across it with Aristotle's critique of the mechanism of Empedocles, to which he objected that in the philosophy of nature the material cause and the final cause *must* both be taken into consideration.

Set forth in the language of modern science, the problem of the origin of living organisms remains as mysterious as it has always been, but its formulation gains in precision. Chemistry, which is the quasi-mechanical form of explanation applicable to the phenomena of life, deals with atoms and molecules. Let us forget about molecules. Properly understood, the passage from atoms to molecules would already require an explanation. Emile Boutroux once wrote on this sort of question a book which is scarcely read anymore but which, although almost one-hundred years old, remains as new as it was in 1874. Let us simply suppose that an explanation of the passage from atom to molecule could be done in a purely mechanical manner. The problem then becomes: How do things move from the most complex molecules to the most simple of living units, the cell? If the cell could be explained in a purely mechanical fashion, there is no reason not to think that the most complex organisms are susceptible of a mechanist explanation, once the cell is given.

In order not to complicate the problem more than is necessary, one ought to refrain from introducing into the discussion the supplemental problem: Are there cells?[20] That is to say, in explaining the genesis of living structures from the most simple elements, which in this case are living cells, is there a scientific justification for affirming that there never has been one or many independent cells [*cellules separées*], capable at least of forming themselves into a living tissue, itself capable of entering into the structure of an organ belonging to some future plant or animal?

This is not a philosophical question. It is simply a question of fact. We ask if anyone has ever seen a *single* living cell, and even if there ever existed a *single* separate cell. Auguste Comte already replied that just as in sociology the individual is an abstraction, so also in biology the organic monads (which he called cells) are abstractions.[21] But there is more to consider: recent attempts to cultivate isolated cells *in vitro* failed. Right up to the present the result of these experiments has been that in order to grow, a living tissue cultivated experimentally "must contain a minimal number of cells, below which cellular multiplication is impossible."[22] And this time we are at grips with a truly important question. Whether it is a scientific question or a philosophical question is a matter of semantics. The question is, admitting that there can be a sound scientific method of explaining nature, to know whether in going from parts to wholes our explanations are not condemned to failure since parts in nature are never given outside of some whole, and, still more, whether the existence of wholes is the final justification of that of their parts? Bergson loved to ask: If I raise my arm, do the positions which it successively occupies in space explain its movement, or does this movement explain the positions which the arm successively occupies in space? So too for living organisms. The whole would not exist without its parts, but is it the parts which have produced the whole, or, rather, is it not that the whole includes the parts as conditions of its own existence? It is impossible to pose these questions without immediately seeing that, in nature such as we see her, no scientific observer has ever seen cells outside of some tissue, nor tissues subsisting spontaneously outside of a living body which itself is a member of a species. These are facts. It is too easy to reserve to science the facts which we can satisfactorily explain and to consign the rest to philosophy. The existence of cells is not contested. The question is only one of knowing whether it is scientifically demonstrated that organisms are "multiples of cells"? If such demonstration exists, we would love to know its whereabouts.[23]

None of these modern biologists mentions the doctrine of final causes. Still more so than "vitalism," "finalism" has become a "dirty word," one to be avoided in respectable scientific conversation. And, nevertheless, the question to which it was a response awaits a response which we wish indeed would be given it.

The only way to find a scientific response to it would be to try to see how the formation of an organism could be explained mechanically not only from cells but also from molecules and atoms. Quantum physics makes the possibility of such a response less likely than ever. We have seen that quantum physics only applies to classes of elements which are completely homogeneous. Such a class alone assures the maximum of predictability in quantum mechanics. Now, living beings are characterized by a maximum of heterogeneity. In the first place, the classes of living beings are not homogeneous, since we never find in them two completely similar individuals. Beyond that, within its class an individual is itself an unhomogeneous substance, since it has a complexity of structure that is almost unlimited.[24] If living matter could be reduced to living cells as its elementary units, the problem would not change. "Even simple cells constitute complex and heterogeneous systems, and the number of different patterns into which one can arrange the vast number of organic molecules, radicals, and electrons that go to make up a tissue the size of a cell is tremendous." Put in other words, the chances of seeing a single living cell come into being from the possible mechanical combinations of its elements are infinitesimal. According to calculations carried out on computers, "there are vastly more such combinations than one could possibly grow cells, even if all the surfaces of all the conceivable planets in the universe were covered with such organisms and were so covered for billions of years."[25]

The only new thing about these ideas is their form. We cannot read these fanciful affirmations without recalling the pages wherein Pascal speaks of the "two infinities," that of greatness and that of smallness. Modern physics simply assists us to see that these truths were still more true than those who discovered them could imagine them to be. Aristotle's objections to the mechanism of Empedocles were far more justified than Aristotle could conceive. In the light of modern science the probabilities that organic structures are spontaneously born from elements mechanically in motion are infinitely small [*faibles*]; so much so that we can say that they do no exist.

What result has Elsasser obtained? He has shown the extreme improbability that living beings could exist in a universe which was solely mechanical. Starting out from pure mechanism, which supposes perfectly homogeneous series of perfectly homogeneous beings, beings as little homogeneous as plants and animals ought not exist. However, they exist. The physicist is content to think that at all events, though infinitely improbable, their existence is not absolutely impossible; but the philosopher who, in this matter, is but the man-in-the-street, remains perplexed. If the existence of such beings is so highly improbable, how has it come about that they exist? And the only response that he can imagine is that it is perhaps necessary to restore to life some ancient forgotten or despised notions. What is to be done, asks G. Canguilhem in his substantial essay on *La Théorie Cellulaire*? He replies:

> It would be absurd to conclude in this regard that there is no difference between science and mythology, between measurement

and dreaming. But, inversely, to desire to radically devaluate ancient intuitions under the pretext of theoretical progress [*dépassement*] is insensibly but inevitably to come to be able no longer to understand how a stupid humanity could become intelligent over time. We do not always dismiss the miraculous as easily as we think we do [*on ne chasse pas toujours le miracle aussi facilement qu'on le croit*], and in order to suppress it in things we at times restore it in thought, where it is no less shocking and, basically, useless.[26]

It is moreover possible to think in this area without mythologizing. Passing his own conclusions in review, Elsasser gives excellent examples of what could be an appropriate discovery of ancient notions in the light of new facts. He calls *organismic* his own response to the problems of life, by which he means that *"organisms represent a separate form of matter."*[27] The proposition is surprising, first of all, because Aristotle indeed believed in the existence of two types of matter, celestial and sublunary; but that notion has been abandoned since Galileo, and here is someone who at the present time asks us to admit the existence in living beings of another kind of matter than that of unorganized beings made from only physicochemical elements! This time it is Aristotle who would have protested, for he had conceived of inorganic matter in such a fashion that, thanks to its substantial form, it could compound with itself in the structure of organized beings. There is perhaps nothing purely material in nature. The mechanistic reformation worked by Descartes first of all demanded the elimination of the philosophical notion of "substantial form"; it is therefore possible to understand how a modern scientist could be led to a conclusion as extraordinary as that. Since there are no more forms in the universe of quantum physics, a specific, even generic, difference between two immense classes of beings can only be explained by a difference of matters. Our biologist sees this clearly, and this it is that leads him to a "more profound, philosophical" view. "No matter how closely it fits the facts of direct observation," any "theory of organisms" will not "be fully satisfactory at a more fundamental level unless it embodies the valid expression of an idea so often emphasized throughout the history of biology, namely, that *organisms* represent a separate form of matter."[28]

Is the idea true though ancient? As much as it reminds me of the old conception, it appears rather new. The notion that living beings are capable of being divided into two parts, one strictly determined by the laws of physics and chemistry, and the other by nature different and autonomous,[29] would have appeared absurd to Descartes and senseless to Aristotle himself. I think the Philosopher would have said: Yes, organic living beings and inorganic things constitute two distinct classes, but it is not that they consist of two different kinds of matter, but rather because their matters are determined by different forms. This is indeed why philosophers have recourse to this notion of form, situated in matter but not itself material, which mechanism naturally does not want to accept at any cost. The Cartesian elimination of the "formal cause" is what makes it necessary to imagine two kinds of matter, as if matter were such that it could admit of an internal principle of distinction. Rather than give the name "form" (or any other name) to that by which living matter differs from nonliving matter, modern physics simply refuses to give it a name at all.

Moreover, without asking ourselves if this new position does not amount to that of the biologist Rostan, whom Claude Bernard denounced for attributing to "organisation" its own efficacy, we welcome this organismic theory[30] and watch its efforts to call to life a certain number of ancient ideas.

Initially Elsasser discerns the presence of an analogy between the scientific notion of "class" and the old philosophical concept of "universals." "Although moderns are more abstract and operational than philosophers of the middle ages, it is not to be wondered at that some of the problems and quandaries connected with inquiries into the nature of organisms are perennial."[31]

Next, Elsasser notes in facts indications of the presence of elements belonging to an order other than the physical. This fact leads him to say that "(by a positivistically descriptive approach one is often able to eliminate implicit metaphysical assumptions). It must appear, in our case, that there is some specific, more or less hidden obstacle which makes this manner of approach less fruitful in biology than it is in physics."[32]

In the third place, we note the distinction between the physical order and the biological order: "we assume that there exist regularities in the realm of organisms whose existence cannot be logico-mathematically derived from the laws of physics, nor can a logico-mathematical contradition be construed between these regularities and the laws of physics. In brief, the existence of such regularities can be neither proved nor disproved on the basis of the laws of physics. Questions regarding the derivation of these regularities from the laws of physics belong to the unanswerable kind."[33]

His fourth notion is that of the presence in living beings of an element which is nondeducible from physics:

> Since, however, physics is valid in the organism, and since life according to the view proffered here can only appear through the inhomogeneity of structures and classes, the concept of the living state has a certain elusive quality which, in a more empirical context, has puzzled all the thinkers of biology. This elusiveness is, however, part of our most common experience. *No biological theory can possibly claim to be taken seriously unless it contains some symbolic representation of this elusiveness in its foundations.*[34]

A fifth notion is that "basic concepts are patterned after the classes of biology, be they trees, cows, or cockroaches. If complexity reaches a certain degree, the common language of everyday intercourse may be more enlightening than a thicket of mathematical formulas." We recognize that this manner of speaking can be astonishing in our times, but, our scientist continues, "given the inhomogeneity of many classes of biology, conceptual language may frequently be the more suitable and appropriate means of expressing the basic relationships pertaining to the regularities of biological theory." In other words, the biomathematical works less well than the physicomathematical.[35]

A sixth notion: Elsasser believes not only in the actual objective reality of biological classes but also in the real existence of hierarchical order, or rather hierarchies of order, between the classes of living beings: "The existence of such hierarchies is a patent fact of immediate biological observation; it is certainly not just an abstract deduction from the analysis of complicated data."[36] Of course, every attempt to formalize mathematically this concept of "hierarchies of order" or of classes is likely to show itself deficient in correspondence to reality and, consequently, as not useful in practice.

These considerations lead Elsasser to one last notion, the name of which at least is familiar to philosophers even if the sense of it remains mysterious to them, that of *individuality*. The new biology conceives it to be a configuration, or a process, which is "an immensely rare occurence if viewed abstractly against an immense number of possible configurations (or processes)."[37]

One immediately feels the fleeting and slippery elements in such a definition. The rarity of a configuration does not explain its nature or even its existence; it is the result of them. It is as organic that a living configuration is rare, and every organic configuration is individual by definition. "Individuality is a universal property of organisms."[38] It "clearly increases as one rises on the evolutionary scale. One might even use individuality broadly as a measure of evolutionary advance."[39] We hail the advent of the notion of evolution in our inquiry. We shall before long meet with it again. For the present let us note that it leads to another very old notion (it goes back to Genesis), a notion which by us is rendered under various forms, but always nourished by that part of fervor with which man speaks voluntarily of himself: at the summit of the scale of evolution stands man. "Man is then the 'highest' of organisms quite simply (what a wonderful adverb!) because men, due to the complexity of their brains, can exhibit a vastly higher degree of individuality than any other kind of organism." Once more we come across a biological fact as little explainable mechanically as it is immediately evident. We shall not be astonished therefore that such a scientist feels himself to be to a certain extent separated "from the more rigorous methods of the physical scientist." Here are his final words: "Anybody who deals with these fields, in addition to being a scientist, should have some capabilities of intuition, perhaps on occasion a little of the poet in him, if he is to apprehend clearly the intricate marvels of Creation."[40]

Creation, with a capital "C" in the original. In French, at least, capitals for other words than proper nouns always make me uneasy. I never know what they signify exactly, and I experience a disagreeable feeling that someone wishes to cause me to take something for somebody. In any case it would not be prudent here to take the word "creation" in its theological or religious or even properly metaphysical sense. Rather, its probable meaning is "the totality of factual reality," the entirety of things that fall under our observation. The principal concern of the new biology appears to be to follow a middle course between vitalism and mechanism, only, in doing so, it brings to our attention the disturbing fact that the very existence of the biological is not susceptible of a mechanist explanation, and that, of course, not only insofar as it exists but insofar as it implies the existence of organized beings. It is because he himself intensely perceives that shortcoming that Elsasser pays attention to statistical physics, in order at least to leave the door open to the possibility of that of which one cannot deny the reality.

The facts that Aristotle's biology wished to explain are still there. He is reproached, sometimes bitterly, with having explained them poorly, but up to the present no one has explained them any better [*plus de tout*].[41] Mechanist interpretations of these facts, which Aristotle formerly said had failed, have not ever been satisfactory; they have only displayed more and more the inevitability of the notions of organization and teleology invoked by Aristotle in order to explain the existence of mechanistic structures of which science is the study. Contemporary science itself attests to the unavoidable necessity of notions of this sort. This fact encourages us not to hold them as no longer applicable, but rather to see in them constants of the philosophy of nature, which itself, within limits accessible to historical observation, does not appear to have ever ceased to be what it is.

CHAPTER VI

The Constants of Biophilosophy

BY BIOPHILOSOPHY, OR philosophy of life, we understand the philosophical interpretation of the characteristic proper to living beings. Life itself is not in question here, nor vitalism, for life is an effect rather than a cause, and vitalism is not a constant in the philosophy of nature. It is not quite true that vitalism, in the sense in which "life" is understood as a distinct energy proper to living beings, the cause of their structure and of their operations, has been professed by all philosophers of nature. Aristotle, we have said, does not call upon "life" as a cause or a principle. It is for him the specific effect of the soul, the idea [*notion*] of which is something else. When we see in finalism a "more subtle and very supple form" of vitalism, we set the discussion on a way which is not the best.[1] The notions of vitalism and of final causality are not necessarily connected.

According to Lemoine finalism admits "that each being is made for its environment, each organ constructed in view of its own function: the vital phenomena tend toward a precise end, from which tendency the name of 'final causes' is derived."[2] There is a composite portrait of finalism which perhaps does not exactly fit any finalist philosophy in particular. In the first place, we can speak of the adaptation of beings to their environments without admitting that they had been "made" with this in view. Next, to conceive of each organism as "constructed in view of" something is to look at the problem from the perspective of the Demiurge of Plato's *Timaeus* or that of the creator God of Judeo-Christian theology. Creationism is not necessarily connected to finalism anymore than vitalism is. Finalism does not even require that the phenomena of life tend toward a "preconceived" end. Whether in fact they do or not is up to the theologians to decide. When the moment arrives for them to search out whether final causes have as their origin divine thoughts and intentions, the philosopher of nature will have decided long ago about their existence on the basis of fact drawn from the observation of nature herself. The biophilosopher is not a theologian.

This mixture of theology and philosophy of nature has exercised a disturbing influence on the history of teleology. Supposing that, as we think, the living world gives witness to the presence of final causality in all the beings which constitute it, and at the same time the theologian, speaking in the name of first philosophy or metaphysics, affirms the existence of a God who is the creator and ordainer of nature: it still remains most often impossible to infer the intentions of the Creator from the inspection of creatures alone. Here it is necessary to allow that Descartes was right when he denied that man could be seated at the council of creation and speak as if the intentions of God were known to him. It is necessary furthermore to grant to the biologist that, side by side with marvelous results, nature abounds with disconcerting failures and flawed workmanship: sickness, the destructive ferocity of beings who live only by the death of others, the colossal mess in the reproduction of plants and animals in which seeds perish in their billions without this prodigality corresponding to any intelligible necessity. If one thinks on the other hand of what ought to be the infinite wisdom of an all-powerful God and compares the detail of his work to his attributes, it is hard to defend him against the feeling that a simple human engineer would easily find many ways to ameliorate the details of his work. We recognize that such problems exist, but they exist for that part of the disciplines of theology and metaphysics which Leibniz called theodicy, or the justification of God against the objections drawn from the existence of evil. It is not necessary to the biophilosopher that natural teleology be perfect in order to authorize him to say that it exists. If it exists, perfect or not, it is the display of nature alone that allows him to decide on the matter.

From this point of view the situation does not appear very different today from what it was in the time of Aristotle. There is still teleology. Basically, everyone speaks as if there were, but, unable to say in what it consists, science prefers to ignore or deny it.

Further, there are beings formed from homogeneous parts and beings from heterogeneous parts. Those which make up the second class are organic beings, made up of parts which are themselves complex and arranged in a certain requisite order if their operations are to be possible. Today we speak of structures, but the structure of a living being does not explain anything; it is the structure which it is necessary to be able to explain. And today, just as in Aristotle's time, it remains impossible to explain how the parts of such a being are arranged, whether it be in themselves or in connection with each other, without the intervention of principles other than those of mechanics. This fact explains why since the time of Aristotle biology has appealed to two complementary principles in order to explain the structure of organic beings, the material cause and the motor cause on the one hand and the final cause on the other. Explanation by means of the material and motor causes already corresponded, in its spirit, to a science of Cartesian type. It presaged modern "reductionism." Explanation by final cause has always been of an entirely different type, in that the principle of explanation which it involves is not in itself the object of empirical observation. The end is not a cause which we can observe at work as we can the motor cause of bodies which collide. For the same reason the end is neither measurable nor calculable; we can only say of it that it exists. On the other hand, we can speak of it with assurance because the effects which we require it to account for are visible, tangible, and perceptible with an obviousness equal to that which we have for extension and movement: they are the very structures of these organic beings. The alteration of order which takes place when we pass from the inorganic to the organic has been well defined by Auguste Comte as the passage from an order in which the parts precondition the whole to an order in which the whole shapes the parts and, in a sense, precedes them. We in our turn shall say that it is as if the parts were there only in view of the whole, or at least as required by it. This is what we call the order of final causality.

The existence of that order and of the relations which it involves is an immediate certitude, although the nature of that certitude remains a mystery to the understanding. It results from a line of reasoning subsequently integrated in perception. We see that a rock is not of the same nature as a tree. However many paving stones we may take from a block of granite, each of them is identical in nature to that of the block: the analysis of one is adequate to describe others and the whole. Organic being is, on the contrary, a whole defined by the aggregation and order of the parts which compose it, and even if its detail escapes us at first sight, we see directly that such an order exists. We see that a being is organic as we see at first glance that some debris we might come across is the remains of a machine or of one of its parts. If the astronauts had come across a plant or animal on the moon, they would have recognized it simply by seeing it. We say that primitives take a watch for an animal, but only the genius of Descartes has been able to take animals for watches.

The spontaneous inference of which we speak is no longer a logical operation composed of explicit judgments, and it may never have been such. It rises rather from psychology, itself understood as the biology of the functions of knowing, which it already was for Aristotle. Its foundation is the perception of beings capable of self-movement. No animal can be mistaken about it. A cat or a dog which looks indifferently at what is before its eyes, a garden for instance, immediately fixes its attention on any moving object: a cat can be fascinated by an infinitely small displacement stirring on a floor or carpet. A number of animals know that "to play dead" is a precaution which is useful in order "not to be seen." Someone out for a stroll, not looking at anything in particular, moves his eyes spontaneously to follow "anything that moves." On the other hand, Aristotle has for some time now drawn attention to the connection between notions of movement and heterogeneous parts, without which self-movement is impossible. The parts of which machines are composed are quite different from those of which living organisms are composed: machine parts are homogeneous in structure, do not know how to substitute one part for another in case of failure, do not reproduce themselves, do not heal themselves, do not produce the energy which moves them. They are no less adjusted to one another than the organs of a living being and "function" effectually like organs of a living body. Machines are artificial imitations of organisms. A man could not fail to note that as soon as he has himself made the most simple instruments and utensils. From this fact stems the quite just observation of Georges Canguilhem that "the vocabulary of animal anatomy, in western science, is rich in names of organs, viscera, segments or regions of the organism expressing metaphors or analogies." Sometimes an organ is designated by the analogy of its role with that of a fabricated item: *sack, aqueduct, axis,* etc. At other times utensils are designated by names taken from organs: *arm, ball-and socket-joint, teeth,* etc. In all cases of this kind, as G. Canguilhem excellently says, "the Greek and Latin denomination of organic forms makes it appear that technical experience communicates certain of its structures to the perception of organic forms," and the reverse is true also.[3] There is no difference between asking oneself what the function of an organ is, that for which "it is useful," and asking what its end is. This is the old problem *de usu partium*. It is a problem as regards organisms and machines, and to pose the problem it is sufficient to perceive either one or the other.

To the extent that we invoke it to give an explanation of this fact, teleology is the object of sensible experience, not in itself but in its effects. This is a question, not of an abnormal or exceptional case but, on the contrary, of one of those numerous cases where in sensible experience itself an immediate inference is produced in the intellect from the perceived effect to the cause. It is quite true that nothing is in the intellect which has not first been in the senses, but neither is anything in the senses of an intelligent being which is not at the same time in the intellect. That can be seen from sensible perception. No one has even seen "dog" or "tree," which are collective classes and not individuals, but we do not cease to perceive splotches of color given shape by the forms which the intellect knows to be vegetable, animal, or human. Likewise for the effects of final causality. There is no essential difference between seeing that a being is organic [*organisé*] and seeing that it is a dog. Intellectual induction from sensible perception is the same in both cases; it is the same case.

From this we draw the immediately obvious conclusion that science has no need for final causes, but it is no less true that what we call final causality exists in reality. The temptation to take this methodological abstraction for a real elimination is perhaps irresistible, but what one has decided not to take into consideration, perhaps even because one has the obligation of averting one's mind from it, does not thereby cease to exist. The explanation of the movement of a traveler seated in a train can be made entirely in terms of mechanism: I pass through a certain distance, at a certain average speed per hour, in a certain time, thanks to the functioning of a machine expending a certain kind and quantity of energy. The mechanist analysis of the situation can go on forever, not only because it involves, beyond the circumstances of my personal life, the immense network of social, economic, and political conditions which a public transportation company disposes, but in the last analysis because such calculation would be theoretically possible. The entirety of this analysis may not, however, answer the question which the traveler might ask himself: What am I doing on this train? For the true response would be: I am going to Marseilles. No scientific method of observation allows the disclosure of the presence in me of that intention (*in-tendere*), whose origin in my thought I perhaps do not know clearly myself. In any case, it is not the intention which transports me, but it utilizes the immense mechanism of the "means" of transportation as if the intention constituted the ultimate justification of those means. It is a thought which utilizes electrical energy, without itself being apparent in the deployment of that energy. The biologist is in a similar situation: he observes, to the exclusion of all teleology, something which could not exist without that teleology, and he has doubtless the right scientifically, and perhaps even the obligation, to do so; but he treats organisms like travelers who would have arrived infallibly at the end of their voyage without having had the intention of going there.

This situation is not perhaps entirely healthy, for it is not certain that the "how" of an operation can be separated from the "why" which is its goal. An exhaustive mechanist explanation of the birth and development of a living being from conception to adulthood would still be an explanation of a process oriented toward a goal which is its *end*. Where there is no end, as in a machine out of control, the process repeats itself indefinitely to the extent of its derangement, and it is then the *"how"* itself which ceases to exist.

If we ask the philosopher What is teleology? it is his turn to be embarrassed. The root of the difficulties of his attempt, if he tries to respond, is perhaps that he tries to define it in itself, as if it were, in the living being, something distinct from that being. The motor cause set aside, for the motor is always distinct from the moved, the causes immanent in the being do not have any other real being than its own. Matter, form, and the end are real constituents of being, but they only exist in it and by it. This is what distinguishes the teleology of nature from that of art. The artist is external to his work; the work of art is consequently external to the art which produces it. The end of living nature is, on the contrary, consubstantial with it. The embryo *is* the law of its own development. It is already of its nature to be what will be later on an adult capable of reproducing itself. Descriptions of natural teleology which situate the cause of it outside of it appear to be conceived in view of justifying the negation of such teleology. At least they deceive themselves as to the object. It could be that the metaphysician and the theologian, in search of a supreme goal of nature, consider themselves justified in positing an Alpha who would also be an Omega as cause and goal of all that is, but the problem which biophilosophy poses to itself is not that. Whatever may be the transcendent origin of it, the teleology of the organism is in it as, once let fly by the archer, that of the arrow which flies to the target without knowing it, is in the arrow. Twenty intentions exterior to it could have directed it toward the target, but it goes there henceforth by itself, and it is indeed the arrow which reaches the goal. The direction of a movement is part of that movement.

Our excellent master André Lalande said one day in the course of his lectures: "Teleology does not admit of being reconstructed." He was right, but it is doubtless because teleology itself is not a construct. It is not mechanism inside out, which Bergson too often reproached it with being. The mechanist thinks that it is a matter of a temporary situation and that since the time of Aristotle no one has found a way of approaching the final soluton of this problem. It is hard to deny this, but it could also be that no scientific way of approaching a problem of this sort exists. The latter possibility is much more likely. The triumphs of mechanism in the recent past will continue in the future insofar as they apply their methods to the order of material, physical realities which consist essentially in extension and movement. We do not know how to determine the limits of this order, but the existence of objects of knowledge whose nature eludes mechanist explanation is no longer an impossibility. This is the case if the order of the immaterial and the unextended is not pure nothingness.

Since he himself is a material being, the knowledge of the scientist is assuredly tied to matter, but it is not matter. We have seen the face of Einstein, but have we seen his knowledge [*savoir*], his thought ceaselessly moving between two or more possible physical universes? We have heard his voice, which was sensible, but how is it that we have perceived the sense of the words he pronounced? If there is that which is knowable, and that which is known, then there is that which is immaterial, and since it is tied to our body, which is sensible, the knowable then exists in the sensible. *There* is a fact which constitutes one of the oldest constants of philosophy. The inevitability of Platonism, in its own right or mediated through Aristotle, comes visibly to the surface here. Since only knowledge could have conceived these things, matter then has the immaterial in it. Centuries, millenia of philosophical speculation have puzzled over the source from whence this immaterial could come. Aristotle replied before them: "From without." Translated: scientifically speaking, we do not know.

Nothing is more out of style than animism; it played an important part in the history of philosophy up to the sixteenth century at least, which was the time of its triumph. Aristotle had conceived of the soul in order to explain the phenomena of life, from the most elementary to the most exalted. It is necessary to recognize once more what he taught, namely, that we only know of the soul through its effects. The various names which we give to it do not give information about what it is in itself: *eidos, morphe, forma* are so many symbols locating the site of an unknown, whose existence is beyond doubt. The name which perhaps might suit it with the least amount of inaccuracy would be the Greek *logos*, translated by the Latin *ratio*, if we would understand these words as a code or intelligible formula of the nature of organic beings, the law immanent in their structure and their development. The only utility which this gives a name is that it warns us to forget its existence, while affirming the thing it names, although we are not able to say what that thing is.

The Cartesian extermination of forms and souls of all sorts is a philosophically irrevocable operation, in the sense that even if we doubt its complete success, we cannot forget that it was tried, and hence it is possible. We shall note, however, that Descartes set aside from the massacre one substantial form, the human soul, of which, contrary to the Aristotelian conception, he attributed to us a direct intuition, not only with respect to its existence but also regarding its essence. Furthermore, it will always be possible to imagine that the operation may have succeeded, for it was logical that it should take place. La Mettrie and many materialists (notably among the Marxists) count Descartes among the number of their ancestors. Biological mechanism and all modern "reductionisms," with their astonishing success, make this reasonable. Final causes have disappeared from science, but have they disappeared from the minds of scientists?

If they are the object of a sort of direct view in their effects, we cannot see how, despite the interdict which keeps them beyond the doorway of laboratories, they would not continue to haunt the minds of scientists. That has been denied, however. "These attempts at explanation," writes a modern biologist, "were driven out of physiology in the first half of the nineteenth century, by those who placed that science on the ground where it stands today. Notable among them was Claude Bernard." To which, furthermore, the same scientist immediately adds, speaking of Lamarck and Darwin, that we have gone so far as to "suspect of latent and, so to say, occult finalism the protagonists of the theories of evolution themselves."

Concerning Lamarck, Darwin, and even the vehement Thomas H. Huxley we have seen that suspicion is an understatement. We cannot maintain that the functions of Lamarck, which produce the organs which they need to function, operate thus *in order to* without operating *in view of* some end. With respect to Darwin we can read the texts wherein he himself speaks explicitly as a finalist, and those other texts wherein his immediate disciples praised him with having reconciled mechanism with teleology. Claude Bernard himself is far from giving evidence in favor of a world of life purged of all final causality.

It is true that Claude Bernard established the science of living matter upon its actual basis, namely, that of experimental physiology. He knew himself to be a born physiologist, so to say, as witnessed by the moving words of his *Cahier de notes*: "Physiology, physiology, it's just part of me [*c'est en moi*]!" However, experimental physiology incarnate kept the pace of a philosopher: "Physics and chemistry only give an account of the performance of a physiological phenomenon, but not of its directive cause, which is by nature living, being a series from the point of departure created by evolution." Something else, then, is necessary in order to extricate ourselves from this philosophical vagueness. But what? Here is Bernard's response:

> *On teleology*: When we see in natural phenomena the enchainment which exists in such a fashion that things appear to be made with foresight of an end, as the eye, the stomach, etc., which form themselves in view of food, future lights, etc., we cannot prevent ourselves from supposing that these things are intentionally made, with a definite end in mind. Because, in effect, when we ourselves make things in this fashion, we say that we make them *with intention*, and we could only admit [as an alternative] that chance has made everything. Well! It would appear that if, when we make things in a fashion in which they concord with a specific end, we say that there is an intentional intelligence of ours, we ought then recognize in the entirety of natural phenomena and their specific connections with specific ends a great intentional intelligence.
>
> This intentional determination appears evident above all in living beings which form a finished whole; it appears less so for the physicist and chemist who only see fragments of general phenomena of the great whole. So the latter are those who have struggled against teleology as furnishing false ideas, and today

> scientists do not dare to avow that they are teleologists because teleology entails things which are not demonstrable. In any case, nothing has been put in teleology's place, and the place remains empty.[4]

After a brief discussion of the hypothesis of the preexistence of germs (Bernard is doubtlessly thinking of Bonnet), then a discussion of adaptation to milieu (he is certainly thinking of Lamarck), Bernard concludes: "Without doubt, we can say all that and many other things still, but these are suppositions, no different from teleology, which is worth just as much as they are until a new order of things appears."

Claude Bernard still sees things then a bit like Aristotle saw them. Like him, he departs from the observation of the workmanlike teleology of man and extends it to the universe of living beings. He states that although physicists and chemists, who do not live in constant contact with vital phenomena, refuse to admit the existence of teleology, or at least refuse to admit that they believe in it without wishing to admit it *because these phenomena are things which are indemonstrable* [*qui ne se démontrent pas*], we have found nothing yet to replace the teleology which we no longer want.

A fanatic of scientism has gone so far as to say that Claude Bernard's texts have been falsified by obscurantists in order to support their own opinion. Things are more simple than that. Claude Bernard knows better than anyone that the life of the individual is evolution;[5] he simply admits what is irresistible in the temptation to think that it is in some manner directed. His role, as a scientist, is not to speculate on the nature of that directive intentionality. In our time, furthermore, summarizing the general conclusions of an inquiry conducted by a group of biologists on the present state of these questions, and thinking moreover of Claude Bernard, its director concluded:

> I think that there exists virtually in nature an infinite number of living forms of which we are ignorant. These living forms could be in some fashion dormant and expectant. They would appear when their conditions of existence present themselves, and, once realized, they would perpetuate themselves as much as their conditions of existence and succession perpetuate themselves.[6]

The situation has thus changed less since Aristotle than one would think, since it is still a question today of "drawing forms from the power of matter" where they are found in potency. The students of Rabelais who, good disciples of Avicenna, hailed their tavernkeeper, "Hey! Giver of Forms!" in order to get another round of drinks were pretty much at home in that science. It is, moreover, man who has become the giver of forms, for the modern biologist already sees himself creating species of living beings not seen until now, of which there is no reason to limit the number. It is always imprudent to set limits to the future of science. Claude Bernard did not do so, but in his time, which is not so far removed from our own, he many times stated that biology did not have any hold on heredity. Today we are rich in knowledge concerning this realm, and the day approaches when the biologist will, on the contrary, dispose of a redoubtable power over living beings yet to be born, whether they be the products of nature or even works of his own invention. H.G. Wells gave proof of a remarkable sobriety of imagination in his *Island of Dr. Moreau*, when we compare his romantic anticipations to the quasi-delirious "scientific" imaginations with which volume V of the *Encyclopédie française*, a work of pure science, draws to a close.[7] Biological finalism bears up quite well, and it does not appear that the fundamental notions which inspire it have changed much at all. Some of our biologists say so with a reassuring candor: "Life is not a phenomenon like other phenomena." This is indeed what Aristotle said some time ago, and for the same reason: "Life involves an organization made up of heterogeneous parts."[8] This heterogeneity appears refractory to all explanation by homogeneous parts as such. It is doubtless destined to remain so.[9]

What happens on the opposite side? It also has its constants, or at least it has one, and since we have followed teleology up to modern times, we ought also to follow its adversary to our times.

We know, moreover, that constant already; it is negative, for mechanism gives no explanation of the existence of its machines. Its scientific fecundity is admirable; it is science itself. But insofar as it claims to resolve the philosophical problem to which finalism is the response, mechanism is a pure nonentity. The sole response at its own disposal is, as we have said, chance, which is not a cause but simply an absence of teleology. Teleology is perhaps a contestable explanation; chance is the pure absence of explanation. We could say that, scientifically speaking, we ignore the question of why birds have wings, but to say that the conjunction of conditions necessary to the flight of birds *was accidental* is to say nothing. To add to chance the astronomical extent of billions of years during which it has been at work is still to say nothing, for whether the absence of a cause lasts a year or billions of years, it remains forever an absence of cause, which, as such, can neither produce nor explain anything.

We remain surprised, then, but also, we must admit, disarmed, in the face of certain professions of mechanist faith, such as that of Julian Huxley, the inheritor of the speculative combativeness of his ancestor Thomas Henry Huxley, whose impetuosity often carries him along to imprudent use of language. For example: natural selection "operates with the aid of time to produce improvements in the machinery of living, and in the process generates results of more than astronomical probability, which could have been achieved in no other way."[10] Here we have an inadvertent comedy, which we can avoid only by saying that, scientifically as well as philosophically, the mechanism of natural selection is simply a nonexplanation.

The [mechanistic] heirs of Lamarckism are not in any more favorable position, for to conceive of the organism as directly shaped by the environment, without the mediation of its needs, confounds the reason as much as it does the imagination. However, in order to arrive at this stage, it is necessary to close one's eyes to the fact that the production of organs through needs may be a cryptofinalism. Discouraged, certain neo-Lamarckians timidly resort to natural selection and to the trick that it operates spontaneously among living beings, but the explanations will not thereby become easy. "Particularly for species of large size, which are generally composed of few representatives, it can be mathematically demonstrated that selection has little to do with it, and that chance plays a particularly significant role in their extinction or survival." Death does not choose intelligently. Thus, then, shall we say, "finding itself actualized through a series of chances, we are tempted to consider the organic world as the result of teleology."[11] One is taken aback by such modest demands concerning intelligibility.

This absence of intellectual rigor is disconcerting when we see it in scientists who are so rigorous in their own scientific research and who appear not to be troubled by it anymore when they undertake to reflect upon their own science. George Gaylord Simpson, professor of vertebrate paleontology at the Museum of Harvard University, considers that to deny evolution "is almost as irrational as to deny gravity."[12] There is no obvious connection between the two cases. The comparison will become valid when the laws of evolution become comparable in precision to those of gravitation and will have been established like them. This is not the case now, and perhaps it never will be, because it is legitimate to doubt that a biological mechanics comparable to celestial mechanics is possible. Philosophers are perhaps wrong in attributing to science a uniformly regular rigor,[13] but scientists could do more to prevent them from being deceived in this regard.

Some adversaries of finalism spend their time in disqualifying it in a plenary fashion by asserting its primary motive to be a religious, almost mystical interest. Sainte-Beuve, who knew his Bacon well, has fortunately described this state of mind in his *Portrait littéraire* of Bernardin de Saint-Pierre: "Final causes never provide a productive view for science, and belong entirely to poetry, morality, religion; at the most they exist only in the scientist's moment of prayer, after which he must return to experiment and analysis."[14]

It is true that propositions of the sort "the pincers of the lobster are made for pinching" are without scientific value. We are nevertheless tempted to say this because they pinch as if they had been made to do so. With a biological scientist we have recalled that they are true pincers. It is not even possible to say that they resemble pincers, for it is our pincers which resemble the first pair of claws of lobsters, spontaneously modified in such a fashion as to become a veritable tool.[15] Here, as in all other cases, it is art that imitates nature, and not the other way around. But this proposition "the lobster's first pair of claws are pincers" has nothing poetic, nor moral, nor religious about it. It is the statement of a fact which no one, be he finalist or not, believer or not, theist or atheist, knows how to contest.

There remains, then, chance as an explanation of the spontaneous growth of these tools: knives, saws, pressure-buttons, etc., which we must exert such calculated ingenuity to procure for ourselves. It has been denied that chance has been invoked as a principle of scientific explanation in biology,[16] but nothing is more true than that it has. In fact, there is no other alternative to final causality: "Finalism encompasses every doctrine which admits that there are facts in the universe which reveal direction."[17] What name should be given to the cause, or to as many causes as you will, the functioning of which reveals no direction? It is true that there are varieties of finalism which are laughably naive,[18] or purely theological, without any connection with science, but a positive notion of teleology appears to be acceptable to some scientists precisely because the opposite view of nature seems unintelligible to them. Such, for example, is the position of the biologist Lucien Cuénot: "Finalist philosophy holds that biology is a special domain . . . ; among its principles (should be listed) the power of invention and organization and the principle of organization which I call anti-chance."[19]

This scientist proceeds prudently. He speaks of "finalist philosophy" rather than science, and rightfully so. Finalist philosophies are responsible to themselves; they do not involve themselves with science at all, and science, as such, has no cause to concern itself with them. The summit of metaphysical finalism was attained by Leibniz, with whom, finalist or not, few other metaphysicians today would agree.[20] On the contrary, we could find few scientists who would not consider that the best explanations, generally, are inspired by the principle that everything happens as if nature proposed to attain certain ends with a strict economy of means.

We will find an illustration of this point in a remarkable episode in the history of science. Maupertuis had been struck by the contradiction between two ways of exploring the phenomena of the refraction of rays of light passing through media of different densities, Descartes' way and Leibniz's way. Maupertuis proposed a novel solution to the problem which reconciled the two points of view by introducing a new principle of explanation, the principle of least action, which is today associated with his name.

The enunciation of the principle implies that an unconscious intention of economy and simplicity of means presides over the laws of nature, which does everything as economically as possible. In other words, of two explanations of the same phenomenon it is likely that the more simple is true. The memoir of April 15, 1744, wherein Maupertuis announced his discovery to the Académie des Sciences, contains an interesting remark:

> I know the repugnance which many mathematicians have for *final causes* applied to physics, and I agree with their repugnance to some extent. I admit that it is perilous to introduce final causes. The error into which men such as Fermat and Leibniz and those following them have fallen only proves how very dangerous their usage is. Nevertheless, we can say that it is not the principle of the thing which has deceived them; it is the precipitateness with which they have taken as the principle that which was only the consequence of it.[21]

Fermat, Leibniz, and Maupertuis himself agreed on the principle of physics, that nature acts by the most simple ways, without useless expenditure, which comes back to saying with Aristotle that nature does nothing in vain, only Maupertuis took thought as to how to apply the rule better. According to him all movement in matter takes place in a fashion such that the *action* required for the course of the movement is as little as possible. This is a principle which the philosopher Aristotle had understood.[22]

It is true that we have left the life sciences in order to enter the territory of mathematics and physics. But mathematics provides science with its most perfect mode of expression, and it also turns out that there is nothing more human than that mathematical formulation of knowledge. Everything takes place in nature by numbers which, however, exist only in the mind of man, the only mathematical animal whom the zoologist comes across in the universe. The more science becomes mathematical, the more anthropomorphic it is, and it is for the scientist a cause of wonderment that the certitude and efficacy of his hold on nature grows in direct proportion as the language of science, itself mathematicized, satisfies more completely the abstract exigencies of his mind. If man is his intellect, and if mathematics is the most perfect manifestation of it, it can be said that the more the knowledge of nature is humanized in virtue of being mathematicized, the more it is useful and true. Thought hesitates on the doorstep of this sort of certitude, the foundation of which escapes it, but of which it can have no doubt.

Compared to generalizations such as the principle of least action, economy of thought, and other similar ones, the notion of natural teleology cuts a modest figure. It can be reproached with being anthropomorphic, but in a science which is the work of man what is not? Furthermore, the important thing is to know whether or not it expresses a fact given in nature, for if we object to final causality as an explanation, it remains as a fact to be explained.[23] It is true that if we make room for it, further problems of a different order than that of natural science and philosophy present themselves. But, first of all, nothing obliges anyone to pose them; and, next, their solutions are not given in advance; and, finally, it would not be reasonable to take exception to so sensible and manifest experience so as to render impossible in advance the posing of certain metaphysical problems, problems that would be susceptible to answers so undesirable [under this scheme] that one might consider it more prudent not to ask them.

APPENDIX I

Linnaeus: Observations on the Three Kingdoms of Nature

CAROLI LINNAEI, Sueci, Doctoris Medicinae, *Systema naturae,* sive Regna tria naturae systematice proposita per Classes, Ordines, Genera et Species.

O Jehova! Quam ampla sunt opera Tua!
Quam ea omnia sapienter fecisti!
Quam plena est terra possessione tua!

(Psalm. CIV, 24.)

Lugduni Batavorum
Theodorum Haak MDCCXXV
Ex Typographia
Joannis Wilhelmi De Groot

Observationes
in
Regna III. Naturae

1. Si opera Dei intueamur, omnibus satis superque patet, viventia singula ex ovo propagari, omneque ovum producere sobolem parenti simillimam. Hinc nullae species novae hodienum producuntur.

2. Ex generatione multiplicantur individua. Hinc major hocce tempore numerus individuorum in unaquaque specie, quam erat primitus.

3. Si hanc individuorum multiplicationem in unaquaque specie retrograde numeremus, modo quo multiplicavimus (2) prorsus

simili, series tandem in *unico parente* desinet, seu parens ille ex unico Hermaphrodito (uti communiter in plantis) seu e duplici, Mare scilicet et Femina (ut in animalibus plerisque) constet.

4. Quum nullae dantur novae species (1); cum simile semper parit sui simile (2); cum unitas in omni specie ordinem ducit (3); necesse est, ut unitatem illam progeneratricem, Enti cuidam Omnipotenti et Omniscio attribuamus, *Deo* nempe, cujus opus *Creatio* audit. Confirmant haec mechanismus, leges, principia, constitutiones et sensationes in omni individuo vivente.

5. Individua sic progenita, in prima et tenerrima aetate, omni prorsus notitia carent, ac omnia sensuum externorum ope ediscere coguntur. Ex *Tactu* consistentiam objectorum primarie ediscunt; *Gustu* particulas fluidas; *Odoratu* volatiles; *Auditu* corporum remotorum tremorem; et demum *Visu* corporum lucidorum figuram; qui ultimus sensus, prae ceteris, maxima voluptate animalia afficit.

6. Si universa intueamur, Tria objecta in conspectum veniunt, uti a) remotissima illa corpora *Caelestia;* b) *Elementa* ubique obvolitantia; c) fixa illa corpora *Naturalia.*

7. In *Tellure nostra*, ex tribus praedictis (6), duo tantum obvia sunt: *Elementa* nempe, quae constituunt; et *Naturalia* illa ex elementis constructa, licet modo, praeter creationem et leges generationis inexplicabili.

8. Naturalia (7) magis sub sensus (5) cadunt quam reliqua omnia (6) sensibusque nostris ubivis obvia sunt. Quaero itaque quamobrem Creator hominem, ejusmodi sensibus (5) et intellectu praeditum, in globum terraqueum locaverit, ubi nihil in sensus incurrebat praeter Naturalia, tam admirando et stupendo mechanismo constructa? anne ob aliam causam, quam ut Observator Artificem ex opere pulcherrimo admiraretur et collaudaret?

9. Omnia quae in usus hominum cedunt, ex Naturalibus hisce cuncta desumuntur; hinc oeconomia mineralis seu Metallurgia; vegetabilis seu Agricultura et Horticultura; Animalis seu Res pecuaria, Venatus, Piscatura. Verbo: fundamentum est omnis Oeconomiae, Opificiorum, Commerciorum, Diaetae, Medicinae, etc. Ex iis homines in statu sano conservantur, a morboso praeservantur, et ab aegroto restituuntur, ita ut delectus horum summe necessarius sit. Hinc (8.9.) necessitas Scientiae naturalis per se patet.

10. Primus est gradus sapientiae res ipsas nosse; quae notitia consistit in vera idaea objectorum; objecta distingumtur et noscuntur ex methodica illorum divisione et convenienti denominatione; adeoque Divisio et Denominatio fundamentum nostrae Scientiae erit.

11. Qui in Scientia nostra Variationes ad Species proprias, Species ad Genera naturalia, Genera ad familias referre nesciunt, et tamen Scientiae hujus Doctores se jactitant, fallunt et falluntur. Omnes enim, qui naturalem vere condiderunt Scientiam, haec tenere debuerunt.

12. Naturalista (Historicus Naturalis) audit, qui partes Corporum Naturalium visu (5) bene distinguit, et omnes has, secundum trinam differentiam, recte describit nominatque. Estque talis Lithologus, Phytologuos vel Zoologus.

13. Scientia Naturalis est divisio ac denominatio illa (10) corporum Naturalium, ab ejusmodi Naturalista (12) judicio instituta.

14. Corpora Naturalia in *Tria Naturae Regna* dividuntur: Lapideum nempe, Vegetabile et Animale.

15. *Lapides* crescunt. *Vegetabilia* crescunt et vivunt. *Animalia* crescunt, vivunt et sentiunt. Hinc limites inter haecce Regna constituta sunt.

16. In hac Scientia describenda et illustranda plurimi omni sua aetate laborarunt; quantum vero jamjam observatum et quantum adhuc restat, curiosus Lustrator facile ipse inveniat.

17. Exhibui heic Conspectum generale Systematis corporum Naturalium, ut Curiosus Lector ope Tabulae hujus Geographicae quasi, sciat, quo iter suum in amplissimius his Regnis dirigat, plures namque Descriptiones addere spatium, tempus, et occasio retardarunt.

18. Methodo nova, maximam partem propriis autopticis observationibus fundata, in singulis partibus usus, probe enim didici paucissimis, observationes quod attinet, facile credendum esse.

19. Si Curiosus Lector fructum aliquem hinc percipiat; illum Celebratissimo in Belgio Botanico D.D. JOH-FRED. GRONOVIO, nec non DNO. ISAC. LAWSON, Doctissimo Scoto, tribuat; illi enim Auctores mihi fuerunt ut brevissimas hasce tabulas et observationes curm Erudito Orbe communicarem.

20. Si comperiar haecce Illustri et Curioso Lectori grata fore, propediem plura, specialioria et magis limata, Botanica imprimis, a me expectabit.

Dabam Lugduni Batavorum.
1735 - Julii 23.

CAROLUS LINNAEUS.
M. D.

APPENDIX II

Darwin in Search of Species

The *Origin of Species* was published in 1859. Its success was not anticipated by either the author or the publisher. Darwin incessantly strove to revise and complete it in successive editions until his death in 1882, at which point the work was in its sixth edition. By 1890 39,000 copies of it had been sold, and who can say to what figure that has grown today, at a time when the work figures in all collections of the great works of humanity, without even considering popular editions of it, hardbound or in paper, and the translation of it into foreign languages.

For a work in so severe a style such success is surprising. In rereading it for a third time, and noting once again how little qualified I was to read it, I came across only two possible explanations for that popularity: either my own exceptional ignorance of geology, paleontology, botany, and zoology did not allow me to appreciate it, or else the remarkable diffusion of the book was due to other than scientific reasons.

I hope that the first reason is the true one, for I am aware of my ignorance, though I ought to have believed it to be more extensive than I realized. Darwin is not only a scientist competent in his specialties but also a man provided with immense scientific erudition due in large part to his own observations, as also to the critical reading which he made of his predecessors and contemporaries. When he describes in detail a flower, the articulation of a bone, or the structure of an insect, he has seen that which he describes. Unless he himself is a competent biologist, his reader has never seen such things. He has not even seen anything of that sort with any frequency and feels no desire to see anything of it. To speak only of myself, it is with secret shame that I read many descriptions of facts which I hear Darwin relating to me, haunted as he is by the impossibility of doing any more than giving samples of his knowledge and his proofs, which he promises to erect into a complete demonstration in a future work. The only contemporaries competent to read him were such scientists as Lyell, Wallace, Huxley, Asa Gray, or Agassiz. Since his death there have been among biologists many readers qualified, at least to some extent, to understand him. But there have been millions of philosophers, theologians, journalists, publicists, and even politicians of every persuasion who have freely discussed Darwin and Darwinism, either as defenders of it or as adversaries of it, without having ever examined a single skeleton or a simple flower.

In order to give a precise focus to my contentions, I shall cite absolutely at random the following paragraph from chapter 3 of the *Origin of Species*:

> Certain compound animals, or zoophytes as they have been termed, called the Polyzoa, are provided with curious organs called avicularia. These differ much in structure in the different species. In the most perfect condition, they curiously resemble the head and beak of a vulture in miniature, seated on a neck and capable of movement, as is likewise the lower jaw or mandible. In one species observed by me all the avicularia on the same branch often moved simultaneously backwards and forwards, with the lower jaw widely open, through an angle of about 90°, in the course of five seconds; and their movement caused the whole polyzoary to tremble. When the jaws are touched with a needle, they seize it so firmly that the branch can thus be shaken.[1]

Darwin's entire doctrine rests upon thousands of facts of this sort, of which he cites but a small sample, and of which most of his ordinary readers have neither any experience nor even any distinct idea. To speak only of philosophers, how many of them are capable of following Darwin's demonstrations based on the branches of the *cirripedia*? And the question here is one not simply of inequalities in degree of knowledge, but of specific differences in intellectual interests. The philosopher only retains from those chapters filled with facts which he reads in Darwin those general conclusions derived from the scientific experience upon which they repose, and which in the thought of the scientist is at the time the sense and justification of that experience. Not only does the philosopher ignore facts, but he even does not wish to know of them when they are described to him. Darwin has probably never led a single philosopher to observe how a bee penetrating a flower of an orchid fertilizes it or how from it others are fertilized. I would almost bet that after having read and reread all that Darwin has said of his dear barnacles, no philosopher ever took the trouble to examine a single one. To ask just how many, of all those who speak of it, have actually read the *Origin of Species* would be an unkind question. It would be better to presume that they speak of it on the basis of hearsay. In any case, even if he has not read and reread Darwin, what a philosopher thinks and says of his work lies in a different order than that in which the thought of Darwin himself does. The properly scientific basis of the doctrine fails to appear in the thought of the philosopher. What he says of it is strictly, as Darwin himself had said of it, *irrelevant*.

The philosopher is struck by Darwin's thought only where the latter, extending beyond the limits of his scientific knowledge, becomes a sort of philosopher without knowing it. Darwin does this often, in a most unself-conscious fashion, and one cannot keep oneself from the impression that his scientific theory itself suffers from these lapses. For it could be said without injustice that when Darwin takes leave of the observation and immediate interpretation of facts, wherein he is the master, he displays an intellectual nonchalance and an imprecision in ideas which does not appear in any way tolerable.

The title of his masterwork was as explicit as possible. It stood thus in the first edition of 1859, and Darwin never changed anything in it: *On the Origin of Species by Means of Natural Selection, or the Preservation of Favored Races in the Struggle for Existence.*

From the beginning a serious imprecision was introduced into the definition of the very object of the book, for at no time did Darwin undertake to clarify the issue of the origin of species in the book, in the sense of the origin of the existence of species. He did not ask himself how it came about that there were species, but rather, given their existence, how it came about that they were such as they were. The problem of the absolute origin of species will never be posed by Darwin. He hardly ever alludes to it even in passing. It will even be observed that, within the limits wherein he poses it, the problem of the origins of the present form of species is not that which he has resolved. In effect, the solution which he proposes for it is the struggle for existence among spontaneous variations, which favors the survival of certain individuals, and, thanks to the hereditary transmission of these favorable characteristics, the progressive formation of a new species. If this is the case, it could be said to him that it is these individual spontaneous variations which are the true origin of species, and it is they, rather than the struggle for existence or the survival of the fittest, which it would be necessary to explain initially. It goes without saying that Darwin was never tempted to do this, and as a result of that there comes about a certain indecision in his purpose.

Supposing it possible to reach some sort of accord on the sense of the word "origin," it remains to define what is understood by "species." Everyone knows in broad outlines what the term signifies: a species is a collection of plants or animals showing traits of resemblance such that it can be distinguished easily from other groups. No one hesitates to distinguish an individual of the species "swallow" from an individual of the species "elephant." The difficulty begins from the moment when, taking any species whatever, one tries to describe the characteristics which define it. Two identical individuals are never found. Considering only those who resemble one another so much that they cannot fail to be classified in one group, one quickly perceives that there exist subgroups, or subspecies, besides varieties which at first are classed within a species, but which next cause one to ask if they are not just as well classified as distinct species. Darwin himself wrote in a state of inextricable perplexity, first dividing a species into varieties, then drawing these back together again as a single species, doing and undoing twenty times the same work without finding a decisive reason for putting an end to the task.

The problem is well known. It could be summed up in the celebrated saying of a modern naturalist: the more one comes to know individuals, the less one finds species. None were more clearly aware of the problem than the predecessors of Darwin in the seventeenth and eighteenth centuries. They are often termed "the classifiers," for their principal problem was to classify living species so as to rediscover the "plan of nature." Species were an absolute necessity for them, and, naturally, fixed species, for what interest could one have in classifying species if they underwent change? For them to talk of classifiable species and to talk of fixed species was the same thing. But since they were acquainted better than anyone with the difficulties of classification, they could not but say, what Aristotle and Buffon have said, that species are only abstract concepts while the sole living realities are individuals.

The attitude of Darwin does not differ essentially from that of his predecessors on this point,[2] except that he is the only one to have written a book on the origin of species, and that it is, consequently, more important for him than for them to know that which is understood to explain the origin. Now, he himself had no difficulty in recognizing that species is only a rather indistinct notion.

> From these remarks it will be seen that I look at the term species as one arbitrarily given, for the sake of convenience, to a set of individuals closely resembling each other, and that it does not essentially differ from the term variety, which is given to less distinct and more fluctuating forms. The term variety, again, in comparison with mere individual differences, is also applied arbitrarily, for convenience' sake.[3]

In the second chapter of the *Origin* there are to be found as many texts as one could wish concerning the impossibility of assigning absolute characteristics by which to distinguish the individual from the variety, and the variety from the species. Darwin vigorously notes the disorder which this incertitude introduces to the classifications, which are nevertheless so necessary to the naturalist if he wishes to know that of which he speaks. But there is little certitude to be had here. "It is certain that many forms, considered by highly competent judges to be varieties, resemble species so completely in character, that they have been thus ranked by other highly-competent judges."[4] It is thus quite difficult to know that of which Darwin intends to explain the origin, unless it be the origin of something which does not exist. It is particularly surprising that the term "species" occupies so visible a place in the title of the work when it plays such a diminished role in the doctrine. A title such as *The Origin of Varieties* would have covered the same problem in its entirety. Groups of similar plants or animals being given, and none deny the existence of such, how is it possible to explain the stability and the fluctuation found in these groups? Are they born of one another, and should we hold their classifications to be so many geneological trees? All that can be discussed without using the term "species," which does not seem to correspond to anything definite.

It could be objected that, in the first place, Darwin does the opposite of this, as the very title of his book shows, and, furthermore, that he even persists in speaking about species in order to say that they do not exist. He had need of the word precisely to be able to deny its existence.

An effort of the imagination is required in order to comprehend the development of his ideas on this problem. All of his predecessors, except Buffon and Lamarck, believed in the existence of species and considered them as fixed. Their position was thus coherent, for a species can be defined only as a class of living beings definable by means of characteristics which are not reducible to those of all other classes. Species is thus by definition a strictly defined type. For it to change would be to cease to be what it is, and thus to cease to exist. To say that species are fixed is a tautology; to say that they change is to say that they do not exist. Why does Darwin so obstinately say that they transform themselves, rather than saying simply that they do not exist?

It is because he never lets his adversaries out of his sight. Perfectly self-consistent, these maintain that since species are fixed, there are no varieties. Let us salute these heroes of logic and mental coherence. If all species are basically nothing more than varieties, why should not all varieties be species? Darwin did not wish it to be so, and the most significant [*dernière*] reason he gives concerning this issue can aid us in clearing up this imbroglio. We are yet in the same second chapter of the *Origin:*

> Some few naturalists maintain that animals never present varieties; but then these same naturalists rank the slightest difference as of specific value; and when the same identical form is met with in two distant countries, or in two geological formations, they believe that two distinct species are hidden under the same dress. The term species thus comes to be a mere useless abstraction, implying and assuming a separate act of creation.[5]

The last remark clarifies the movement of Darwin's thought in this matter, although it may not have been as clear to him, perhaps, as it is to us today. He perceived at least confusedly the propriety with which the mind links together the notions of species, fixity, and divine creation. True or false, the position of Linnaeus and Buffon was clear: species exist, and they are fixed, because in the beginning God created them such as they are even to our day.[6] Darwin knows that a connection exists in the thought of his adversaries between the notion of the fixity of species and that of creation, but he is less philosophical than Lamarck and does not see clearly that the two notions do not have any necessary connection. He thus tries his best to pulverize the notion of species into an indiscriminable multitude of varieties, because *if there are no species, it is not possible that there have been separate creations of them.*

The criticism of the notion of species thus occupies an important place in Darwin's doctrine, and chapter 2 of the *Origin* is particularly instructive in this regard, for one hears there many an echo of the uncertainties through which Darwin himself passed while serving his apprenticeship as a naturalist. At that time at least, one certainty in fact dwelt within him: a road leads from the individual to series of varieties more and more stable and distinct, which themselves lead to subspecies and finally to species, these latter becoming evident only at the end of numerous accumulated variations, in the absence of which the transitions would be always perceptible: "Certainly, no clear line of demarcation has as yet been drawn between species and sub-species – that is, the forms which in the opinion of some naturalists come very near to, but do not quite arrive at, the rank of species: or, again, between subspecies and well-marked varieties, or between lesser varieties and individual differences."[7]

One understands by means of this passage that Darwin attaches great importance to individual differences which are of so little interest to classifiers, for these initial differences, most often infinitesimal [*infime*], are the very points of departure for changes which lead to future species. But one sees at the same time how indeterminate the idea of species remains in his mind. It occurs to him to say in the same phrase that their existence is certain, even though one does not know how to define them. For example, in chapter 4, "Natural Selection":

> In the first place, varieties, even strongly-marked ones, though having somewhat of the character of species – as is shown by the hopeless doubts in many cases how to rank them – yet certainly differ far less from each other than do good and distinct species.[8]

Every attentive reader of Darwin is familiar with the expressions *good species* or *true species*, the good species implicitly contrasted with the bad, the true species with the false. Darwin speaks of this issue again in chapter 9: "It is all-important to remember that naturalists have no golden rule by which to distinguish species and varieties." It is necessary to rely on the judgment of those who are well acquainted with the class in question. Above all, "they grant some little variability to each species, but when they meet with a somewhat greater amount of difference between any two forms, they rank both as species, unless they are enabled to connect them together by the closest intermediate gradations."[9] When Darwin assures us, to reassure himself, that good species *certainly* differ among themselves more than do varieties, his *certainly* is thus a "certainly" of incertitude. Let him who has never made use of such tactics throw the first stone at him! To excuse this, however, does not dispense with the duty of noting the extreme indeterminacy of the notion of species in a work which sets out to explain the origin of species. "Hence in determining whether a form should be ranked as a species or variety," Darwin simply says, "the opinion of naturalists having sound judgment and wide experience seems the only guide to follow."[10] The word is obviously there only by force of habit. Scientifically speaking, it corresponds to nothing.

The profound tendency of Darwin to destroy the species, the study of which comprises the object of his book, appears nowhere better than in chapter 9,[11] on "Hybridism." It has been known since Aristotle that the sterility of hybrids is an obvious sign that the male and the female from which they proceed belong to different species. In other words, two species are really distinct when their crosses are sterile. Darwin is careful not to deny such evidence, but he criticizes it. It appears that this fact disturbs him a bit. "The fertility of varieties, that is of forms known or believed to be descended from common parents, when crossed, and likewise the fertility of their mongrel offspring, is, with reference to my theory, of equal importance with the sterility of species; for it seems to make a broad and clear distinction between varieties and species."[12] It is thus because their crosses are genetically sterile that the most strongly differentiated varieties merit the name of species. By means of an alarming paradox this theory of the transformation of species establishes initially that their genetic fixity is the most evident mark of their reality.

It is understandable that Darwin may have felt some embarrassment about this, but he contrived to diminish as much as possible the import of the fact. It is here that the simple philosopher feels himself incapable of following his arguments concerning causation, for they rest upon facts which, for him, amount to the words which the naturalist uses. The general movement of his argument is nevertheless perceptible, for it appears to lead to the conclusion that even if there is sterility, it is not the abstraction called species which is the cause of it.

Two observers of great experience, Kolreuter and Gartner, experimenting on the same group of plants, in order to establish by the fecundity or sterility of their crosses whether they were varieties or species, arrived at diametrically opposed results. In reality, "neither sterility nor fertility affords any certain distinction between species and varieties. The evidence from this source graduates away, and is doubtful in the same degree as is the evidence derived from other constitutional and structured differences."[13] Let us then make an act of confidence in the numerous arguments accumulated by the limitless erudition of Darwin and designed to show that the sterility of hybrids does not pertain to the fact that they are crosses of species. Presently a third highly qualified horticulturalist will be cited who assures us that his crosses of perfectly pure species show themselves fertile (as if their fertility did not pass normally as certain proof that they were only varieties). At another time Darwin emphasizes cases wherein species can be hybridized and remain fertile more easily than they can fertilize themselves! Darwin is invincible in each particular case, except, perhaps, by one of his peers; but the complete process of reasoning proceeds from a common basis of uncertainty. How could it be certain that the sterility of hybrids "is an extremely general result; but that it cannot, under our present state of knowledge, be considered as absolutely universal,"[14] since the distinction between species and varieties is not perfectly reliable [*sûre*]? Here one begins to feel that the scientist is in fact an advocate who pleads a cause. His mind entertains a favorable prejudice for the fertility of hybrids despite the extreme generality of the contrary position. Everything which can contribute to reduce the stability of species is grist to Darwin's mill. One asks oneself more and more, as one follows further his demonstration, why he continues to speak about it.

For it is in no way necessary to his subject. In a sense, all plants are varieties of the vegetable kingdom, and all so-called species of animals are varieties of the animal kingdom, but it is quite necessary that there should exist species in the ordinary sense of the term if one wishes to be able to prove that they were not created just as they are from the beginning. This, which is Darwin's dominant purpose, thus comes to grips with the problem of explaining how, without having been created as such, species, subspecies, and varieties have been able to form themselves.

The response to the question consists in the law of the struggle for life, also called the law of the survival of the fittest, the idea of which we have seen came to Darwin in the course of reading an essay of Malthus', which Darwin picked up simply as a distraction, and wherein however he found his path. The law of natural selection, to which he held as to the apple of his eye, allowed him to explain how, rendered more fit to survive by the happy accident of favorable individual variations, and transmitted by heredity from generation to generation, certain specific ancient forms were able to be gradually replaced by new forms. In one of his flashes of intuition, where ideas mutually fertilized one another and seemed to fall into place, all the experience previously acquired by Darwin in the realm of the breeding of domestic species came to be seen as a model for the explanation of the formation of species from other species. Spaniels, bassets, and greyhounds hardly resemble each other at all; an English race horse is quite different from a Percheron; yet all are equally dogs and horses. Why, in favoring incessantly the more fit, should not natural selection produce the same diversity in the same unit?

In formulating the question it suffices to see what differences supervene when one responds in the affirmative. It is not a question of knowing whether stockbreeders obtain new species or simply new varieties. At the point in the discussion where we are, it is possible to admit that this distinction is empty. The real difficulty is in knowing who replaces the absent stockbreeder in the natural transformation of species. Darwin's response is well known: it is natural selection which conducts the operation. But the objection thereto is no less well known: How is it that an accumulation of small spontaneous variations can sum up and organize themselves in a certain direction, in such a fashion as to produce the infinitely complex structures of living beings and their organs? To bring to mind only one celebrated example, Darwin himself said that when he posed to himself the question, the thought of the development of the eye gave him chills down his spine. Nevertheless, he maintained intrepidly right to the end that if one should take into consideration the immense extent of the geological epochs and the unimaginable numbers of individuals upon whom nature followed out her experiments, the spontaneous and progressive formation of living beings from elementary and most simple forms, perhaps even from any sort of living matter whatsoever, this could not be held to be impossible.

In the last pages of the *Origin* Darwin courageously confronts Moses and Cuvier at one and the same time, firmly maintaining that "species are produced and exterminated by slowly acting and still existing causes, and not by miraculous acts of creation."[15] Without doubt, "authors of the highest eminence seem to be fully satisfied with the view that each species has been independently created," but he himself prefers "view[ing] all beings not as special creations, but as the lineal descendents of some few beings which lived long before the first bed of the Cambrian system was deposited."[16] In this view these beings appear to him to be ennobled.[17]

One may not know how to deny the grandeur of this view, so somber and even tragic, according to which "from the war of nature, from famine and death, the most exalted object which we are capable of conceiving, namely, the production of the higher animals, directly follows."[18] It is even necessary to acknowledge that it would be vain to undertake to refute this view, whether it be from Darwin's point of view or anyone else's. Like the distinct creation of species – a theological doctrine which he proceeds against with active detestation, without even asking himself on what so-called revealed authority it is founded – this progressive formation of living beings which may proceed by itself "whilst this planet has gone cycling on according to the fixed law of gravity" is a simple mental construction [*une simple vue de l'esprit*] whose merit is to give an explanation in a satisfactory manner (if it is true) of a literally innumerable multitude of facts, observed or observable, present, past, or even future. The totality of universal history is beheld here under a single and simple human glance. One can imagine that Darwin had been enraptured by it, but this is simply to replace one theology by another, and both together are equally indemonstrable. It is possible, moreover, to suspect it. A sober scientific truth is capable of arousing admiration, perhaps even enthusiasm, but of an intellectual variety rather than that sort of popular cult of which, under the name of Evolutionism (which is a stranger to it), Natural Selection has become the object.

Those to whom the method of the great scholastics is familiar find themselves on familiar ground here, in the precise order of rational explanation and in the depiction of the difference of objects. Thomas Aquinas, for example, thirsted for rational explications, but he knew that whatever is in itself the object of religious faith will always escape finally from a completely satisfying rational explication. He thought it at least possible to do two things in favor of the object of his faith: to show that it did not contain any rational impossibility, strictly speaking, that is to say, nothing self-contradictory; then to refute the objections directed against these truths [of religion] and to show thus that these are no proofs, or that they are false. When he speaks of science, Darwin proposes nothing which, in itself, might be inaccessible to reason. Quite on the contrary, the arrangement of living beings within each class in species, genera, and families, which naturalists call the system of nature, is an object eminently intelligible and satisfying for the mind. That which is not so is the manner of conceiving and explaining the origin of it.

Near the beginning of the fourteenth chapter[19] of the *Origin* Darwin notes that it is remarkable in itself, this possibility of arranging in a sort of system all living beings: "The ingenuity and utility of this system are indisputable. But many naturalists think that something more is meant by the Natural System; they believe that it reveals the plan of the Creator."[20] Darwin adds here a caution which perfectly characterizes the foundation of his thought, namely [*savoir*], that to say of the system of nature that it reveals the plan of the Creator "[adds] nothing . . . to our knowledge."[21] What he desires to know, insofar as he is a scientist, is the natural cause and the law which have presided over the formation of the hierarchy of beings according to this plan. Darwin appears to want to say that even if it could be demonstrated that this plan had been willed by God, one could not always know how God had willed that things happen in order to constitute this system of nature. Darwin's own profound intention is precisely to disclose the natural law according to which, created or not, the system is constituted. In other words, it could be said that the magnificent *Systema Naturae* of Linnaeus, whose restrained [*sobre*] tables one cannot take in hand without emotion, is, for Darwin, less a conclusion than a starting point, a questionable response. Only, like the theologian who questions himself on *The Truth of the Catholic Faith*, Darwin sets himself a question which does not allow of a scientific response, not, this time, in theory [*droit*], but in fact. If the system of nature was created, no one knows how it was created. If the Creator simply created those conditions necessary for nature's making of itself, no one knows how it made itself. We say: no one has demonstrable scientific knowledge of the manner in which the living world constituted itself. No one knows moreover whether a God made it or whether it constituted itself completely by itself.

When a firm conviction is not vigorously demonstrated, whatever its nature be, it must be argued. This is what Thomas Aquinas called "to ascribe probable reasons" in favor of belief. Darwin displayed a remarkable power of invention in order to persuade his reader of the truth of natural selection. Even in his purely scientific discussion of particular points he comes to say: "I am not able to see any great difficulty which would prevent this from being the effect of natural selection." But the ignorant ought not give himself over to ridicule by criticizing arguments which are strictly scientific. He should be content therefore to examine the attitude of Darwin upon that which could well be called the critical point of his doctrine. Darwin himself has said repeatedly that he thought of natural selection as analogous to artificial selection. New species are born from nature like they are born in stockbreeding, except that in nature there is no stock-breeder.

In the moments when he did not think about the difficulty, Darwin did nothing to facilitate the solution of it. He did not minimize the importance of the role of the stockbreeder or the lucidity of his calculations. Speaking about what stockbreeders had done for the sheep, Lord Somerville said: "It would seem as if they had chalked out upon a wall a form perfect in itself, and then had given it existence."[22] A Platonic demiurge working, his eyes fixed upon the ideas, could not do better. But it seems obvious that if one does away with the demiurge, the stockbreeder, and the idea, it becomes difficult to explain the birth of such a form.

Darwin knows this, but one hesitates concerning the attitude that should be attributed to him when confronted by this difficulty.

In a passage in the *Origin*, unique so far as I have been able to ascertain, he appears to think that since stockbreeders are able to create new forms, then a fortiori nature is capable of doing so. In chapter 4[23] devoted to "Natural Selection," after having given several of the reasons which explain the armament and ornament of the males in certain species, Darwin makes this remark to himself: "It may appear childish to attribute any effect to such apparently weak means: I cannot here enter on the necessary details; but if man can in a short time give beauty and an elegant carriage to his bantams, according to his standard of beauty, I can see no good reason to doubt that female birds, by selecting during thousands of generations the most melodious or beautiful males, according to their standard of beauty, might produce a marked effect."[24] This brief appeal to the notion of sexual selection, which he will develop at length elsewhere, has nothing in it to affront reason, for there is involved in this case at least the choice of a conscient animal, a spontaneous preference for perceived and known qualities. The immense class of females, acting collectively during millenia, play here the role of the stockbreeder. But how could one explain the choice required in the birth of a new species, when it is a question of favoring the hereditary transmission of *infinite* physiological modifications which are favorable to the survival of the species?

Darwin thought perhaps to introduce there, in the same chapter 4,[25] a previously unforeseen distinction between two kinds of artificial selection, *methodical* selection and that form he called *unconscious selection.* Therein lay new difficulties.[26]

Methodical selection, practiced by stockbreeders and horticulturalists, is pursued with the express intention of producing new varieties. This should be admitted without discussion, more especially as Darwin does nothing to minimize the qualities requisite of a good stockbreeder. He often observed them and admired them. These men have an astonishing quickness of eye, for in this area it is not sufficient for success to be able to separate varieties clearly distinct and have them reproduce. It is necessary to know how to observe the effect "produced by the accumulation in one direction, during successive generations, of differences absolutely inappreciable by an uneducated eye," differences, Darwin adds, which "I for one have vainly attempted to appreciate."[27] Thus it is that in Chapter I, where he wishes to emphasize the importance of "Variation under Domestication," Darwin uses the strongest language at his disposal in order to exalt the role of the stockbreeder: "not one man in a thousand has accuracy of eye and judgment sufficient to become an eminent breeder."[28] Yet this does not say enough: "If gifted with these qualities, and he studies his subject for years, and devotes his lifetime to it with indomitable perseverance, he will succeed, and may make great improvements; if he wants any of these qualities, he will assuredly fail."[29]

It is thus quite a rare bird which presides over the success of methodical selection directed by man. But in chapter 4, where Darwin wants to persuade us that simple animal breeders obtain results comparable to those of nature, working like her, except for her blindness, guesswork, and lack of calculation, it is no longer a question of rare gifts. One asks oneself even how the selection could operate effectively. The notion of unconscious selection is itself little precise. It essentially reduces itself to that of a choice which operates by the intervention of the stockbreeders, and nevertheless by itself as well, since the breeders choose by virtue of a sort of spontaneous and natural flair without a distinct idea of that which they perhaps are going to obtain. In chapter 4, on "Natural Selection," the expression recurs many times: "As man can produce, and certainly has produced, a great result by his methodical and unconscious means of selection, what may not Natural Selection effect?"[30]

It is obvious that the question is far from embarrassing Darwin. However, his courageous affirmation cannot pass as a response, nor can it exempt him from making one. What can nature do in the absence of all conscious selection (for Darwin has said and repeated that in the case of nature *selection* is a simple metaphor), and if the expression "unconscious selection" reveals itself upon examination as completely metaphorical and arbitrary also? For in the case of the progressive evolution of a stem, a shell, or a bone, one can say, if one will, that everything happens as if it had a choice in the matter; but there is no choice involved. Females choose males, but leaves, roots, or bones do not choose at all. One finds oneself reduced, then, to explaining chance as directed by an immense accumulation of sheer chances, each of which taken separately is but the absence of explanation, the regular articulation of which remains enigmatic. One does not know how to demonstrate that this is *impossible*, but one can at least observe that the affirmation of it is totally arbitrary and is only justified by the previous refusal of all other forms of explanation.

It is not possible that Darwin, who reflected so long on this problem, had not perceived what a radical difference there is between speaking of selection in connection with living beings endowed with consciousness, and thus capable of preferring one thing to another, and in connection with things, living or not, devoid of all consciousness of organic modifications of which they are the subject. Darwin acts as if he had not perceived the difference, not without, however, leaving open to chance the rise of the awareness of the distance which separates the two cases.

It pleases Darwin to speak of selection unconsciously operating through the stockbreeders, who spontaneously and randomly [*au juger*] choose the most interesting individual variations to preserve and propagate. But even if they are not conscious of thus preparing the birth of a new species, they are perfectly conscious of working by choice. Its very justification does not completely escape them: "Man selects only for his own good: Nature only for that of the being which she tends."[31] Let us grant him that; it still is the case that man truly chooses between variations which favor, hinder, or make no difference: "He often begins his selection by some half-monstrous form; or at least by some modification prominent enough to catch the eye or to be plainly useful to him."[32] One sees nothing definite in so-called natural selection which takes the place of this choice, but arguing as if he responded to the question,

Darwin persists in saying that since it takes place over a longer time than human selection, the absence of all choice on the part of nature ought to lead to results much more remarkable. This man, ordinarily so calm, becomes lyrical on this issue:

> How fleeting are the wishes and efforts of man! how short his time! and consequently how poor will be his results, compared with those accumulated by nature during whole geological periods! Can we wonder, then, that Nature's productions should be far 'truer' in character than man's productions; that they should be infinitely better adapted to the most complex conditions of life, and should plainly bear the stamp of far higher workmanship?
>
> It may metaphorically be said that natural selection is daily and hourly scrutinizing, throughout the world, the slightest variations: rejecting those that are bad, preserving and adding up all that are good; silently and insensibly working, *whenever and wherever opportunity offers*, at the improvement of each organic being in relation to its organic and inorganic conditions of life.[33]

The enthusiasm of Darwin for natural selection is justified, if it exists; and it exists for him incontestably.

The further one continues to read, the further is he surprised at Darwin's certitude, or, rather, at his absence of inquietude in passing from natural selection to artificial selection.[34] Nor is he totally unaware of the problem. Darwin observes that "sexual selection, by always allowing the victor to breed, [proceeds in nearly the same manner] . . . as does the brutal cock-fighter by the careful selection of his best cocks." This is hardly compatible with unconscious selection.[35] It is so, nevertheless, in the eyes of Darwin, who speaks a bit further on of "that unconscious selection which follows from each man trying to keep the best dogs without any thought of modifying the breed." Thus it stood in the text of the first edition of 1859. Later on, in the sixth and final revised edition which was published in 1872, instead of asserting "*by that unconscious selection*," he added this appropriately inspired revision: "*by that kind of unconscious selection*."[36] This was the furthest that he may have gone toward having a clear conscience in the matter. It still was not to go far enough, for a choice that is not professionally systematic is not necessarily unconscious thereby. When the shell of a barnacle [*coquille d'une patelle*] changes, the modification is truly unconscious, and if this modification is the point of departure for the formation of a new species, the selection which produces it is truly unconscious. Nothing is more unlike this than the process of any sort of human choice whatsoever. It is strictly contradictory to speak, as Darwin does in this same fourth chapter, of the "results of *unconscious* selection by man, which depends [*sic*] on *the preservation of all the more or less valuable individuals*, and on the destruction of the worst."[37] It is quite true that often that may not be done "scientifically" or even "methodically," but it is absolutely not the case that it may be done "unconsciously."

Why is Darwin so attached to that adverb? I do not know of any text where he may have said why he is attached to it. Perhaps he himself was not clearly conscious of the reason for his insistence on the use of a term which one comes to see he knew to be inexact. Without having the right to affirm it, I am inwardly persuaded that Darwin found in the use of this word of sort of alibi. He was indeed conscious, we know, of the colossal extrapolation from selection by domestication implied by the hypothesis of natural selection. The principal argument in his favor was that if it were true, it explained so many things! But, as Claude Bernard said, to explain is not to prove. In default of proving it, one had an impressive confirmation of it if one could imagine that, at bottom, the artificial selection practiced by stockbreeders since ancient times was only one particular form of natural selection. If one can describe artificial selection as being as unconscious as natural selection, the latter benefits forthwith from the quasi-experimental certitude which we have of the former. For this to be the case, it is necessary that nonscientific artificial selection should be unconscious; therefore it is.

Of what ruses are intimate certitudes of all sorts not capable in order to present themselves to the mind as truths objectively founded on reality? Darwin is infinitely attractive; none can gainsay this. He is incontestably a scientist by blood, and quite deserving of respect. One does not sense this less, despite some amusement, when one follows word by word the artifices to which he has recourse at times, and of which he is the first victim. Everyone can discover them for himself in coming across a phrase such as the following, from chapter 4:

> In the case of methodical selection, a breeder selects for some definite object, and if the individuals be allowed freely to intercross, his work will completely fail. But when *many men,*

> *without intending to alter the breed, have a nearly common standard of perfection, and all try to procure and breed from the best animals*, improvement surely but slowly follows from this unconscious process of selection, notwithstanding that there is no separation of selected individuals.[38]

Who does not see that Darwin prepares here an arrangement of things appropriate to his own design? One sets out from a great number of men in order to assure the impersonality of the event. These men operate, as nature does, without the intention of modifying the breed. They are found, moreover, to have spontaneously in common the same ideal of the breed to be produced to guide their operations, and since this breed will result from a concourse of spontaneously harmonized efforts, it will be a product as natural as those of natural selection. Finally, unconscious as that of nature, all these processes will, like nature's, result in the formation of a species more perfect than that whose place it is taking. As if it proceeded by itself to a conceptual crossbreed of great merit, he attributes to art the unconsciousness of nature in order to be able to attribute to nature a "polity" as lucid as that of art. The words are those of Darwin himself: "*nature . . . in its polity*"; "*the polity of nature*"; "*the natural polity.*" Is this language truly that of science? But how much scientific knowledge is in the most exacting thought in matters scientific? Instead of trying to make us take as scientific truths the long train of reveries over which their imagination dallies, scientists would render us the greatest service by warning us as precisely as possible, each time, of the point where their thought, impatient of the rigors of proof, grants itself the pleasure of intelligently imagining what it no longer hopes to know. But perhaps it is necessary to imagine much, in order to know a little.

Notes

Chapter I. Aristotelian Prologue

1. At the beginning of his treatise *On the Parts of Animals* Aristotle distinguishes the properly scientific knowledge of an object from the knowledge which a simply cultivated man, a philosopher, for example, can and ought to have. A good intellectual formation ought to allow us to value correctly the quality of the method followed by any scientist [*savant*] in setting forth the contents of his own science. General culture is that of a man capable of correctly forming judgments of this sort in almost all the branches of knowledge. All questions concerning the order and method to be followed in setting forth a science are within the competence of the specialist of the science. The result of the Aristotelian *paideia* is to confer, in each branch of knowledge, the aptitude to form competent judgments on its object and upon the appropriate fashion of setting it forth.

2. Aristotle, *History of Animals*, I, 1, in *The Works of Aristotle*, vol. 2, Great Books of the Western World, vol. 9 (Chicago: Encyclopaedia Britannica, 1952), p. 7. [Aristotle will hereafter be cited by the title of the work followed by the page and column number of the Berlin Greek text as used in the Oxford translation. All footnotes added by the translator and all other editorial matter inserted in Gilson's footnotes are set off by brackets.]

3. *On the Parts of Animals*, 640^{a}.

4. Ibid.

5. *On the Parts of Animals*, I, 1.

6. Ibid.

7. *History of Animals*, 491^{a}.

8. [*On the Parts of Animals*, I, 1, 640^{a}.]

9. [Ibid.]

10. [Ibid.]

11. [See ibid., I, 1, 642^{a}, 639^{b}.]

12. Ibid., I, 1, 639^{b}. Cf. J. Owens, "Teleology of Nature in Aristotle," *The Monist* 52 (1968): 159-73. [Ogle translates the passage thus, despite Gilson's objection (see p. 3, above) that Aristotle never used "an abstract expression such as 'final cause'."]

13. *On the Parts of Animals*, I, 1, 640^{b}. [This does not agree in all particulars with Gilson's French.]

14. One will find a vigorous and pertinent defense of Aristotle against the reproach of anthropocentrism in Michel-Pierre Lerner, *La notion de finalité chez Aristote* (Paris: P.U.F., 1969). The author refers to A. Mansion, *Introduction à la physique aristotélicienne*, 2nd ed. (Louvain, 1945), pp. 261-62. It is entirely true to think, with M.-P. Lerner, that "to say that nature makes or searches out the best in all things, does not signify for Aristotle that she is related to some demiurge endowed with the faculty of deliberation." On the contrary, she operates as an artist so perfect that she has no need of deliberation in order to attain her end infallibly. As an elite marksman, nature hits the target without the necessity of taking aim. But this is not the question. The wrong sort of anthropomorphism (there are, besides, other forms of it) is to conceive the teleology of nature after the model of that of the artisan, but it is legitimate to infer by analogy from the existence of teleology in the artisan's operations to that which the operations of nature give testimony to. In the two cases there is a manifest adaptation of means to ends. Aristotle says so expressly: "If there is teleology in art, there is finality in nature." In *Physics*, II, 8, Aristotle already uses [*tient*] the two alternate terms which still present themselves in biophilosophy: finality or chance. In his *Historical Sketch of the Progress of Opinion on the Origin of Species* Darwin justly noted, in connection with this chapter of Aristotle: " 'Wheresoever, therefore, all things together (that is all the parts of one whole) happened like as if they were made for the sake of something, these were preserved, having been appropriately constituted by an internal spontaneity; and whatsoever things were not thus constituted, perished, and still perish.' [*Physics*, II, 8, 198[b]] We see here the principle of natural selection shadowed forth." [Charles Darwin, *On the Origin of Species* (New York: Modern Library, n.d.), p. 3, fn.] The last remark would be true if Darwin had not already [*plutôt*], in order to explain the survival of the fittest, counted on a series of chances which would produce the same results as final causality. But the remarks which precede this conclusion are just: Whether it is a question of nature or the artisan, there is teleology each time that a regular and constant series of terms results in, always or most often, the same final term. If nature engendered houses, they would develop as architects construct them; but nature does not construct them. [This is confusing. The "remarks which precede this conclusion " are Aristotle's, though one cannot tell this in Gilson's footnote. The last material in the footnote is a paraphrase of *Physics*, II, 8, 199[a].]

15. Pierre Dieterlen (in *Critique*, no. 246, p. 953, note 1) defines imposture: to affirm as a demonstrated truth an indemonstrable expression and to spread it about to a public which does not know in what demonstration consists. The definition is quite good, and the occasions upon which it may be applied are not rare. Its author gives it in connection with another case than that of final causality.

16. *On the Parts of Animals*, I, 1, [640[b]].

17. [Ibid., I, 1, 641[a]. This is almost certainly the passage which Gilson

seems to be paraphrasing, although he gives it an initial (but no terminal) quotation mark. He makes no specific citation for the passage.]

18. Ibid., I, 1, [641[b]].

Chapter II. The Mechanist Objection

1. Aristotle, *On the Parts of Animals*, I, 1. On the problem itself see Lucien Cuénot, *Invention et finalité en biologie* (Paris: Flammarion, 1941). The greatest defect of this book is that it is throughout so thoroughly reasonable. Thus it could not satisfy any party, and partisan positions are those which receive publicity. See particularly "Deuxième partie: *Le méchanicisme*," pp. 50ff. [sv]. One will notice that geocentrism was an astronomical error of which science was the judge. Anthropocentrism is a philosophical and theological thesis, independent of geocentrism, which is not susceptible to either scientific verification or refutation. Under its pure theological form, moreover, anthropocentrism is connected to theocentrism. If God created the universe for man, and man for himself, then the final cause of the existence of the universe is God, who wished to associate other beings with his glory and his beatitude. These questions have only one point in common with our problem: Is there or is there not teleology in nature? If it does not exist there, then these questions cannot even be asked. If it does exist in nature, then they can be asked, but they are not the questions which we pose. We restrict ourselves to the first purely philosophical point, which is enough for us. It is even enough for us to prove rigorously that it is of its nature not a scientific, but a philosophical, conception. From thence the problem passes into the hands of the theologians.

2. On this aspect of the thought of Descartes see our *Etudes sur le rôle de la pensée médiévale dans la formation du système cartésien*, Etudes de philosophie médiévale, XIII (Paris: Librarie Philosophique J. Vrin, 1951). On the materialism of d'Holbach and his mechanism see E. Gilson and T. Langan *Modern Philosophy* (New York: Random House, 1963), pp. 533-34.

3. After having recalled how mathematics had degenerated in sliding from the speculative to the practical, Plutarch adds: "But what with Plato's indignation at it, and his invectives against it as the mere corruption and annihilation of the one good of geometry, which was thus shamefully turning its back upon the unembodied objects of pure intelligence to recur to sensation, and to ask help (not to be obtained without base supervisions [sic] and depravation) from matter; so it was that mechanics came to be separated from geometry, and, repudiated and neglected by philosophers, took its place as a military art." Plutarch, *Vie de Marcellus*, XXI (éd. Pléiade), p. 680. [See *Plutarch: The Lives of the Noble Grecians and Romans* (New York: Modern Library, n.d.), "Marcellus," p. 376.] Cf. art. XXVII: "Yet Archimedes . . . though these inventions had now obtained him the renown of more than human sagacity, . . . would not deign to leave behind him any commentary or writing on such subjects; but repudiating as sordid and ignoble the whole

trade of engineering, and every sort of art that lends itself to mere use and profit, he placed his whole affection and ambition in those purer speculations where there can be no reference to the vulgar needs of life; studies the superiority of which to all others is unquestioned, and in which the only doubt can be whether the beauty and grandeur of the subjects examined, or the precision and cogency of the methods and means of proof, most deserve our admiration." Ibid., p. 683 [*Lives*, p. 378].

4. Descartes, *Discours de la méthode*, VI (éd. J. Vrin, 1966), pp. 127-28. [See *Discourse on the Method of Rightly Conducting the Reason*, in Elizabeth S. Haldane and G.R.T. Ross, trans., *The Philosophical Works of Descartes* (Cambridge: Cambridge University Press, 1973), I, 119.] If one compares this with the text of Plutarch just cited in the previous note, it could be said that Descartes is here the anti-Archimedes. One asks whether Descartes did not have the intention of setting himself in opposition to Plutarch in writing these lines of the *Discourse.*

5. Francis Bacon, *On the Proficience and Advancement of Learning Divine and Humane*, II, 7, 3. Knowledgeable concerning scholasticism, Bacon remarks pertinently that it is vain for its masters to attach so much importance to the knowledge of formal causes, since, properly understood, they are unknown to us. Always solicitous to conserve the old terminology to the greatest extent possible, Bacon will continue to speak of "forms," but in the new sense (and mechanist spirit) of the "latent schematism" hidden in things. The true *forms* of things are their *laws*. Physical science takes into consideration all natures, but solely with respect to their material and efficient causes, and not with respect to their forms (op. cit., II, 7, 5). Since the end is the form, physical science will thus abstain from taking into consideration either the form of things or their ends.

6. Aristotle, *On the Parts of Animals*, I, 5 [Gilson erroneously cites I, 1], 645[a].

7. P.A.M. Dirac, "The Evolution of the Physicists' Picture of Nature," *Scientific American* 208 (1963): 47.

8. [Darwin, *Origin* (M. L. ed.), p. 38. Not cited by Gilson.]

9. F. Bacon, *On the Proficience*, II, 7, 3. The most profound objection directed by Bacon against the search for final causes is that according to the scholastics themselves this search is tied to that of the substantial forms, and the latter are unknown to us. In effect, *if* practical efficacy is the end of knowledge, the argument is irrefutable. It only remains to determine if, in fact, such causes exist in reality, unknowable in themselves, but recognizable through their effects. For that which concerns Bacon himself, see op. cit., II, 7, 5.

10. F. Bacon, *Novum Organum*, I, 51. [See E. A. Burtt, ed., *The English Philosophers from Bacon to Mill* (New York: Modern Library, 1939), p. 38.]

11. F. Bacon, *Advancement of Learning*, II, 7, 6.

12. [Francis Bacon, *Of the Proficience and Advancement of Learning,*

Divine and Moral, in *The Works of Lord Bacon* (London: Bohn, 1871), vol. 1, p. 37.]

13. An old gardener said to me one day: "It is the leaf that makes the grape." We intentionally refuse to take into consideration arguments against final causality which mechanism draws from monsters and from imperfections of whatever observable nature in the structure of living beings. See Lucien Cuénot, *Invention et finalité en biologie*, pp. 58-85. Cases of lack of development [*d'atélie*], hyperdevelopment [*hypertélie*], and improper development [*dystélie*] presume a situation in which one can identify proper development [*eutélie*]. There are no monsters except in relation to normal beings. Finally, and above all, it is not a question of knowing if natural teleology is universal and perfect, but if it exists. The problem which its imperfections pose is, in part, that of the ill-formed [*mal physique*] with which the theologian ought to concern himself, but its discussion is not incumbent upon the philosopher of nature in general or of biophilosophy in particular. There enters into the mechanist mentality a quite strong dose of unconscious anthropomorphism: if I had created a living being, I would have done it better. We can see faults of fabrication which could be avoided. The proportion of seeds to the number of living beings who attain maturity reveals a frightful mess, and so on. Imperfections observable in teleology do not prove its nonexistence any more than imperfections and miscarriages in a machine authorize the hypothesis that it made itself without engineer or workers.

14. F. Bacon, *Advancement of Learning*, II, 7, 7.

15. Aristotle, *On the Parts of Animals*, I, 1. [640^{b}. What Aristotle says (in Ogle's translation) in the last passage cited, though, is that "the formal nature is of greater importance than the material nature."]

16. R. Lenoble, *Mersenne ou la naissance du mécanisme* (Paris: Librairie Philosphique J. Vrin, 1943; 2nd ed., Paris, 1971). For the citation which follows see p. 3.

17. I. Newton, *Optics*, III, 1, 28 [in Great Books of the Western World (Chicago: Encyclopaedia Britannica, 1952), vol. 34, pp. 528-29].

18. Ibid., p. 529. Newton's conclusion goes beyond the limits of what our contemporaries would allow to be called science or even natural philosophy: "and these things being properly ordered, do not phenomena show us that there is an incorporeal Being, living, intelligent, omni-present, who in infinite space (as in his *sensorium*) sees things themselves intimately, knows them completely, and thinks? In this philosophy, each step forward does not give us immediately perhaps the knowledge of the First cause, but it brings us nearer it, and, for inasmuch as it does so, it ought to be for us the highest prize." Aristotle would have approved this conclusion, and, further still, Thomas Aquinas: *Summa contra gentiles*, IV, 1.

19. [C. Bernard, *Leçons sur les phénomènes de la vie communs aux animaux et aux végétaux*] Republished by Georges Canguilhem (Paris: Librairie Philosophique J. Vrin, 1966). On the sum of these problems see G.

Canguilhem, *La connaissance de la vie*, 2nd ed. (Paris: Librairie Philosophique J. Vrin, 1967).

20. C. Bernard, *Leçons sur les phénomènes*, p. 31.
21. Ibid.
22. Ibid., p. 32-33.
23. Ibid., p. 206.
24. Ibid., p. 292.
25. Ibid., p. 293.
26. Ibid., p. 336.
27. Ibid., p. 338.
28. Ibid.
29. Ibid., p. 397.
30. Ibid., p. 344.
31. Ibid., p. 370.
32. Jean Rostand, *Les grands courants de la biologie* (Paris: Gallimard, 1951), p. 198. Cf. T. A. Goudge, *The Ascent of Life* (Toronto: University of Toronto Press, 1961), p. 131.

Chapter III. Finality and Evolution: A. Fixism

1. "In the natural order, the perfect precedes the imperfect, as act precedes potency." *Summa theologiae*, I, 94, 3, resp.
2. "Nature proceeds from the imperfect to the perfect in all things begotten." Ibid., I, 101, 2, "sed contra."
3. "In affirming whatever may be, we ought then to follow nature, except for those things which depend on divine authority, which is above nature." Ibid., I, 99, 1, resp. The phrase is manifestly inspired by some known formulas of St. Augustine: what we know, we owe to reason; what we believe, we owe to faith.
4. Descartes, *The Principles of Philosophy*, part III, ch. I, 45, 46.
5. [In the French edition, Gilson refers the reader to "Appendix I," which contains the Latin for that part of the title Gilson uses, as well as the Latin for all twenty of the "Observations on the Three Kingdoms of Nature" of the Leyden edition dated July 23, 1735. Since Gilson appears to use a variant edition, a facsimile of the Latin original is included in this edition as Appendix I. See M.S.J. Engel-Ledeboer and H. Engel, trans., *Carolus Linnaeus systema naturae, 1735.* Facsimile of the first edition with an introduction and a first English translation of the "observationes" (Nieuwkoop: B. de Graaf, 1964). The translation used here is on pp. 17-19 of the Engel-Ledeboer and Engel translation.]
6. C. Linnaeus, *Fundamenta botanica, in quibus theoria botanices aphoristice traditur*, 2nd ed., augmented: *Philosophie botanica*, aphorism 132. There is a French translation of this work: *Philosophie botanique* (Paris and Rouen, 1788). It has not been available to me [Gilson].

7. *Fundamenta botanica*, V. Sexus, aph. 132: "Initio rerum, ex omni specie viventium (3), unicum sexus par creatum fuisse suadet ratio." Cf. Descartes cited above: "and natural reason persuades us." I admit to some difficulty in translating *sexus par*; this "sexual pair" appears to be simply a male and female couple.

8. Buffon, *Oeuvres philosophiques*, ed. Jean Piveteau (Paris: P.U.F., 1954), p. 355. Buffon concludes that "an ass is an ass, and not a degenerate horse, a horse with a hairless tail." The example removes all possible doubt about the sense of "degeneration"; it is truly a degeneration [*dégénérescence*]. Cf., further on, note 15, the other curious example alleged by Buffon, and moreover in Buffon's work itself the whole chapter "On the Degeneration of Animals." One asks oneself if the shade of original sin does not haunt his zoology.

9. Buffon, "Premier discours" to the *Histoire naturelle générale et particulière*, in Piveteau, ed., op. cit., p. 10. [Gilson has made two slight and inconsiderable changes in the text.] Buffon adds: "This truth is too important for me not to press everything which could render it clear and evident." An example from botany: "anomalous plants whose species is in the middle between two genera," and so on. "This pretension which botanists have of establishing perfect and methodical general systems is thus ill-founded," p. 10. The point is aimed mainly at Linnaeus, the prince of "classifiers." The classification of animals into six classes by Linnaeus is quite arbitrary and quite incomplete, p. 18. It has been known since Aristotle that the mare [*jument*] has no mammary glands (sic), p. 19. [Buffon's French reads: "cependant depuis Aristote on sait que le *cheval* n'a point de mamelles," p. 19 (emphasis mine). Gilson simply complicates an already murky situation by changing "cheval" to "jument." See John Lyon, "The 'Initial Discourse' to Buffon's *Histoire naturelle*," *Journal of the History of Biology* 9, 1 (Spring 1976), "Introduction," p. 139 and n. 14.] Buffon further on proceeds to a harsh criticism of Linnaeus, mixed with praise for the ancients, "I do not say in physics, but in the natural history of animals and minerals," p. 20.

10. Buffon, "Premier discours," in Piveteau, ed., op. cit., p. 19.

11. "Nature is the system of laws established by the Creator for the existence of things and for the succession of beings," *Histoire naturelle*, ed. cit., p. 31. [Gilson's French does not quite follow Buffon's in the edition cited, for Gilson has "pour la création des êtres," while Buffon wrote "pour la succession des êtres."] It is not one thing (which would be everything) nor one being (which would be God), but a "living power" which animates everything and is subordinate to God, op. cit., *Invocation à Dieu* ["Première vue," *Histoire naturelle*], p. 35.

12. "An individual, of whatever species it may be, is nothing in the universe, a hundred individuals, a thousand, are still nothing; species are the sole beings of nature – perpetual beings as old, as permanent as she; and in order to judge better we should not consider them anymore as a collection or a series of similar individuals, but as an entity [*un tout*], independent of

number and time, an entity always living, always the same; an entity which has been counted as a unit in the works of creation and which consequently is but a unity in nature," op. cit., p. 35. Permanence of species, p. 38; but all the individuals are different, p. 38; fixism, p. 289.

13. "Boethius, amused, attends the process, hearing the one side and the other spoken of competently, but failing to see what ought to be accorded to each, he does not claim to solve the conflict definitively." It is still not solved. Never has the reality of species been so much contested as since it has been taught that they undergo transformation. "From whence comes the idea of species? Evidently from practical necessity. Man indeed has to designate beings which he knows and separates from other beings by a particular name." Lucien Cuénot, *Encyclopédie française*, t. V, 18-1. The author is thinking here of the hunter, the fisher, the farmer, perhaps also of the naturalist bent upon classifying. But the practical necessity would be without object if there were no species. The fisherman with a line has need to distinguish the gudgeon from the roach and the perch only because there are fishes belonging to different species. It remains moreover true that according to Deslongchamps' saying, "the more individuals we have, the fewer species" (op.cit., V, 18-2).

14. Buffon, *General History of Animals*, ch. I, ed. cit., p. 236. This view goes back to Aristotle. "Nevertheless, if he were not able to define and, especially, name species, Aristotle indeed saw the essential character of it, the same that we use as a criterion, that of reproduction. . . . There, then, we have species defined by coupling and fecundity, absolutely as it is in our day." Edm. Perrier, *La philosophie zoologique avant Darwin* (Paris: Alcan, 1884), p. 13. If the possibility of fecund interbreeding defines a species, its existence only becomes certain from the moment when it is impossible for it to evolve.

15. In the same chapter on "The Ass": if the ass and the horse came from the same stock, if they were of the same *family*, one could draw them closer together again, remake horses with mules, and "undo with time what time may have done." Let us think about this pearl: if they could not reproduce together, "the Negro would be to man what the ass is to the horse." ["L'Ane," *Histoire naturelle des animaux*, ed. cit., p. 357. [What Buffon says is not quite what Gilson has. Buffon writes: "si le Nègre et le Blanc ne pouvoient produire ensemble . . . il y auroit alors deux espèces bien distinctes; le Nègre seroit à l'homme ce que l'âne est au cheval, ou plutôt, si le Blanc étoit homme, le Nègre ne seroit plus un homme, ce seroit un animal à part comme le singe." Gilson wrote: "Recueillons cette perle: s'ils ne pouvaient se reproduire ensemble, 'le nègre serait à l'homme ce que l'âne est au cheval'."]

16. This passage from the chapter on "The Ass" is judiciously called to our attention in Edm. Perrier, op.cit., p. 61. [The original is in the chapter on "The Ass" from the *Natural History of Animals*, Piveteau ed., p. 355. Gilson must be citing the version in Perrier, for his French does not exactly follow Piveteau's version. I have tried to indicate some of the differences by bracketing the material from the Piveteau edition.]

Chapter III. Finality and Evolution: B. Transformism

1. Lamarck

1. [Gilson cites from Lamarck, *Philosophie zoologique*, nouvelle édition, by Charles Martins, 2 vols. (Paris: F. Savy, 1873). There is an English translation of this work by Hugh Elliot, which we shall use for the passages Gilson cites: *Zoological Philosophy* (London: Macmillan, 1914).]

2. Cf. "Trying to persuade by reasoning rather than by positive facts, Lamarck shared the eccentricities of the German philosophers of nature: Goethe, Oken, Carus, Steffens. Today, we reason less and demonstrate to more purpose." Charles Martins, "Introduction biographique," ed. cit., t. I, p. vii. The fact might perhaps also be explained principally by Lamarck's belonging to the tradition of Diderot and the generation of the *Encyclopédie.* He himself was the author of four volumes of the *Encyclopédie méthodique.* These several lines from the conclusion [of the *Philosophie zoologique*] will give the tone of the times: "Nature, that immense ensemble of various beings and bodies, in all the parts of which there dwells an eternal cycle of movements and changes which the laws govern, the whole alone unchanging, as it has pleased its sublime author to make it exist, ought to be considered as a whole constituted by its parts for an end that its author alone knows, and not exclusively for the ends of any of its parts. Each part necessarily having to change and cease to be in order to make up another part has an interest contrary to that of the whole; and, if it should reason, would find the whole poorly formed. In reality, however, the whole is perfect and completely accomplishes the end for which it is destined." *Philosophie zoologique*, t. II, p. 426. [I do not find this passage in the "Summary" of part 2; nor is there any "conclusion" to the work in Elliot's translation. Gilson is quoting from the "Additions Relative aux Chapitres VII et VIII de la Première Partie," volume and page as just cited.]

3. Lamarck, *Philosophie zoologique*, ed. cit., t. II, p. 248 [Elliot, *Zoological Philosophy*, "Table of Contents," x]. "Now the true principle to be noted in all this is as follows:

(1) Every fairly considerable and permanent alteration in the environment of any race of animals works as a real alteration in the needs [*besoins*] of that race.

(2) Every change in the needs [*besoins*] of animals necessitates new activities on their part for the satisfaction of those needs, and hence new habits.

(3) Every new need [*besoin*], necessitating new activities for its satisfaction requires the animal, either to make more frequent use of some of its parts which it previously used less, and thus greatly to develop and enlarge them, or else to make use of entirely new parts, to which the needs have imperceptibly given birth by efforts of its inner feeling; this I shall shortly prove by means of known facts." Lamarck, *Philosophie zoologique*, 1re partie, ch. VII [Elliot, p. 112]. – "It is not the organs, that is to say, the nature and shape of

the parts of an animal's body, that have given rise to its special habits and faculties; but it is, on the contrary, its habits, mode of life and environment that have in course of time controlled the shape of its body, the number and state of its organs and, lastly, the faculties which it possesses." *Ibid.* (sic) [Elliot, p. 114. Lamarck is here quoting a passage from his earlier work, *Recherches sur les corps vivants*, p. 50 (sic).]

This notion of the progressive production of the scale of living beings from elementarily simple generations is the only coherent view that can be given to transformism. Referring to his own *Esquisse d'une histoire de la biologie,* Jean Rostand writes: "The fundamental idea of transformism, that is to say the idea of the formation of the complex from the less complex, the superior from the inferior." ("Les précurseurs français de Charles Darwin," in the *Revue d'histoire des sciences et de leurs applications*, 1960, pp. 46–47.) This idea in effect appears to be the only one which is common to all transformisms. It is met with, moreover, under this pure form only in the philosophy of Spencer.

4. Lamarck, op. cit., I, ch. 3; t. I, p. 2 [Elliot, p. 35]. "It has been imagined that every species is invariable and as old as nature, and that it was specially created by the Supreme Author of all existing things." Op. cit., I, 3; t. I, p. 74 [Elliot, p. 36].

5. There is no ground for thinking that this notion had been suggested to Darwin by Lamarck. Every naturalist ascertaining an element of variability in species finds himself *ipso facto* in opposition to Linnaeus and Buffon on this point. Nevertheless, this notion took on a vital importance with Darwin which it never had with Lamarck.

6. "But these groupings . . . are altogether artificial, as also are the divisions and subdivisions which they present. Let me repeat that nothing of the kind is to be found in nature, notwithstanding that justification which they appear to derive from certain apparently isolated portions of the natural series with which we are acquainted. We may, therefore, rest assured that among her productions nature has not really formed either classes, orders, families, genera or constant species, but only individuals who succeed one another and resemble those from which they spring. Now these individuals belong to infinitely diversified races, which blend together every variety of form and degree of organisation, and this is maintained by each without variation, so long as no cause of change acts upon them." *Philosophie zoologique*, 1re partie, ch. I; t. I, p. 41 [Elliot, pp. 20–21]. A sort of principle of specific inertia happily prepares things for a biology of mechanist inspiration.

7. Op. cit., I, 1; p. 43 [Elliot, p. 22].

8. Op. cit., I, 3; p. 71 [Elliot, p. 35. Here, however, the phrase is cast in the conditional; "or if they have not in the course of time" etc.]. "Species . . . have only a relative constancy and are only invariable temporarily." Op. cit., ch. III, t. I, p. 90 [Elliot, p. 44]. Lamarck's gaucherie of style is only too perceptible in this "temporary invariability" of species, but we see what he wants to say and it is not necessary to abuse him for it.

9. Op. cit., I, 3; t. I, pp. 75–77 [Elliot, p. 37].

10. It is because the duration of human life is minimal compared to that of the intervals between the great changes undergone by the surface of the earth that species appear to us to be stable. Op. cit., I, 3; t. I, p. 88 [Elliot, pp. 42–43].

11. Op. cit., I, 3; t. I, p. 90 [Elliot, p. 44]. Moreover, Lamarck offers this definition as having practical value: "to facilitate the study and knowledge of so many different bodies" [Ibid.].

12. Lamarck, *Philosophie zoologique,* I, 7; t. I, p. 224 [Elliot, p. 107. The first passage above ("the environment affects") is in italics in the English].

13. Op. cit., t. I, pp. 235–38 [Elliot, pp. 112–14].

14. Lamarck cites in this connection a passage from his *Recherches sur les corps vivants*, p. 50, where he established the following proposition: "It is not the organs, that is to say, the nature and shape of the parts of an animal's body, that have given rise to its special habits and faculties; but it is, on the contrary, its habits, mode of life and environment that have in the course of time controlled the shape of its body, the number and state of its organs and, lastly, the faculties which it possesses." Cited in *Philosophie zoologique*, I, 7; t. 1, pp. 237–38 [Elliot, p. 114].

15. *Philosophie zoologique*, I, 7; t. 1, pp. 248–49 [Elliot, p. 119].

16. [Indeed he has not. The ludicrous material here is to be found in Elliot, pp. 119–20, almost immediately after the previous citation. Gilson, however, cites] *Eloge de M. de Lamarck*, by M. Cuvier, in the *Mémoires de l'Académie royale des Sciences de l'Institut de France*, t. XIII (Paris, 1835). "It is not the organs, that is to say the nature and the form of the parts, which give rise to habits and faculties; it is the habits, the manner of living, which, with time, give birth to the organs; it is by means of wanting to swim that the membranes of the feet of water birds are developed; by means of going to the water, by dint of wishing not to get wet, that the legs of the waterside bird are elongated; by means of wanting to fly that the arms of all [involved] turn into wings, and fur and scales into feathers; and that it should not be thought that we add or cut out anything, we use the author's own words." Op. cit., xix–xx. Yes. The spiteful tone alone is Cuvier's. We recognize nevertheless that this *éloge* does not deserve the ill repute that it has attained. It was difficult to conceal Lamarck's numerous scientific misadventures in many areas. But Cuvier did not fail to situate Lamarck's greatness where it belongs: his invention of the class of "animals without vertebrae"; and he honored as he ought the heroic grandeur of the man, his indomitable courage, his passion for work, and all that with a bare minimum of income which often confined him to misery.

17. Cuvier, *Eloge*, p. xx.

18. Cuvier, op. cit., pp. xx–xxi: "Everyone can perceive that independently of many paralogisms of detail [the explanation] also rests on two arbitrary explanations: one is that it is the seminal vapor that organizes the embryo; the other, that desires, strivings, can engender organs. A system resting on

such bases can tickle the imagination of a poet; a metaphysician can derive from it a completely different generation of systems; but it cannot sustain for a moment the examination of anyone who has dissected a hand, viscera, or only a feather."

19. Lamarck, *Philosophie zoologique*, I, 3; t. I, pp. 74–75 [Elliot, p. 362].

20. "In the production of living bodies, both animal and plant, nature *was originally obliged to create* the simplest organisation. . . . *She soon had to* endow these bodies with the faculty of multiplying, for otherwise she *would everywhere have been occupied* with creations, and this is beyond her power. . . . Now . . . , *she hit upon the plan*. . . . Such is the method employed by nature for the multiplication of those animals and plants." etc. *Philosophie zoologique*, II, 8; t. II, p. 138 [Elliot, p. 275. There are some slight divergences from the French here. The passage is to be found in part II, ch. 9, not ch. 8, as Gilson says.].

21. Op. cit., II, 9 [Gilson gives II, 8]; t. II, p. 151 [Elliot, pp. 280–81].

22. Op. cit., I, 7; t. I, p. 230[a] [Elliot, p. 113: "The structure of animals is always in perfect adaptation to their functions."].

23. [Elliot, p. 120.]

24. [Ibid.]

25. Op. cit., I, 7; t. I, pp. 254–55 [Elliot, pp. 121–22]. Cf. in the same chapter, p. 256, the effects produced on the body of the kangaroo because it carries its little ones in the pouch it has under its abdomen [Elliot, p. 123]. On the hereditary transmission of character thus acquired, pp. 258–59 [Elliot, p. 124].

26. "If for a being adaptation to an environment resides in the fact that it acquires from it advantageous characters, is this not a finalist solution – as that of natural selection is – capable of accentuating this advantageous particularity?" Paul Lemoine, *Encyclopédie française*, "Les êtres vivants" (Paris, 1937), t. V, 08–2. And, a little further on the same page: "To be sure, the word 'adaptation', which we use so facilely, covers over a 'frightening question'. But the phenomenon, severed from the simplistic aspect given it by Lamarck, is undeniable, and the cases in which it appears in a convincing manner are numerous." The word is frightening because it is only another way of saying "teleology." Adaptation is still a respectable term, scientifically speaking; it is a way of not becoming accused of finalism.

27. Lamarck, *Philosophie zoologique*, ed. cit., I, 7; t. I, p. 263 [Elliot, p. 126]. For Darwin's harsh judgments of Lamarck see Jean Rostand, "Les précurseurs français de Charles Darwin," in *Revue d'histoire des sciences et de leurs applications* (P.U.F., 1960), p. 54. Cournot's remarkable text (1857), cited by M. Jean Rostand (p. 57), probably would on the contrary have found favor in Darwin's eyes.

2. Darwin without Evolution

1. Darwin will be cited from Charles Darwin, *The Origin of Species and the Descent of Man*, vol. 49 of the collection The Great Books of the Western

World, edited by R. M. Hutchins and M. J. Adler, Encyclopaedia Britannica, The University of Chicago, 1952. A work useful for following the changes introduced by Darwin in [the various editions of] his first great work is *The Origin of Species: A Variorum Text*, edited by Morse Peckham (Philadelphia: University of Pennsylvania Press; 1959). In the *Origin* the "Glossary of the Principal Scientific Terms Used in the Present Volume," prepared by W. S. Dallas at Darwin's request (Peckham, pp. 761–71 [pp. 245–57]), does not contain the word "evolution." The "alphabetical list of Chapter Sub-Titles" (Peckham, pp. 797–99) and the general index based on the sixth edition (Peckham, pp. 801–16) do not contain the word either. Finally, and perhaps most importantly, the final recapitulation, so important as evidence of the scientific terminology of Darwin himself (Peckham, pp. 747–59), does not contain the word "evolution" in any of its numerous variations. The theory of natural selection (Art. 183·1·b) dominates these pages (p. 748), not to the exclusion of other views, but in order to define that of Darwin. [Hereafter the Great Books edition of the *Origin* used by Gilson will simply be cited as *Origin*, followed by the appropriate page numbers. For the sake of ready reference, the citation in the Modern Library edition of the *Origin* (New York: n.d.) is given in parentheses after the Great Books citation.]

2. I ask pardon for translating thus the English title of Darwin's work *The Descent of Man*. I do not propose to alter the usage, or even to contest it, but it appears to me to be so ambiguous that I allow myself the privilege of not submitting to it. In English, *descent* signifies in the first place the act or fact of descending, then extraction, or origin, and finally, lineage. In French, *descendance* signifies, first connection of filiation, or posterity: *une nombreuse descendance*. In this sense the "descendance" of man would be the "Superman" of Nietzsche, or Jules Romains' *"unanime."* One can, properly understood, use words in any sense that one wishes, so long as one defines that sense. Thus, one can call "descendance de l'homme" the act by which man is descended from . . . , as if one spoke of the descendance of a staircase, rather than the descent of a staircase. But we do not have any intention of reforming usage. For the author of *The Origin of Species*, for whom the problem of the origin of man is but one particular instance, it is indeed a question of the series of biological events which lead from a certain species of primate to the human species. It is therefore of the descent of man, and not of his descendance, that Darwin means to speak. But we make no accusations; we excuse ourselves.

3. "Now things are wholly changed, and almost every naturalist admits the great principle of evolution." This whole passage will be translated later on in this chapter [p. 97, French ed.; p. 96 above]. We shall speak in this chapter primarily of Darwin as a biologist, and insofar as his transformist views are of interest to the problem of finality. Only incidentally will the theological problem be spoken of, whose importance however, for a certain time, dominated his mind. On this point see pp. 91–93 [French edition; pp. 92–94 above]. For that time it is the man who is involved, rather than the scientist.

4. Charles Bonnet (of Geneva, 1720–1793), *Oeuvres complètes*, t. VII (Neuchatel, 1783). See the "Tableau des Considérations sur les Etres Organisés," particularly chs. XIII–XVII, pp. 61–72; and *Palingénésie philosophique ou idées sur l'état passé et l'état futur des êtres vivants*, pp. 113–60 (particularly part III, ch. IV, "Préformation et évolution des êtres vivants," pp. 151–55). For the passages which we cite, see pp. 263–65. On Bonnet, see *Mémoires autobiographiques de Ch. Bonnet, de Genève*, by R. Savioz (Paris: Librairie Philosophique J. Vrin, 1948). R. Savioz, *La philosophie de Charles Bonnet, de Genève* (Paris: Librairie Philosophique J. Vrin, 1948). H. Daudin, *De Linné à Jussieu* (Paris: F. Alcan, n.d.), pp. 101–5 and 161–63.

5. M. Serres, *Principes d'embryogénie, de zoogénie et de tératogénie* (Paris: Mémoires de l'Académie des Sciences, 1860), t. XXV: 940 pp., with plates. Serres cites Bonnet in ch. II, pp. 20–21, and the whole book is directed against him: "There are no true metamorphoses in nature, the system of preexistences concludes (Bonnet, *Corps organisés*, p. 44); everything metamorphoses in the realm of organized beings, replies epigenesis. On which side is truth found, with two so contradictory assertions? On the one hand is absolute fixity, immobility, death; on the other, movement, change, life. Repose on one side, movement on the other: such is the contrast of the Old and the New Testaments of the natural sciences from the point of view of the science of growth [*développements*]." Op. cit., ch. VII, pp. 75–76. The theory of *epigenesis* goes back to W. Harvey, *Exercitationes de generatione animalium*, exercise 51, where Harvey declares himself in accord with Aristotle on this point: *"Per epigenesim sive partium super exorientium additamentum pullum fabricari certum est."* Cf. T. Huxley, art. "Evolution" (in biology), *Encyclopaedia Britannica*, 9th ed. (New York, 1882), vol. VIII, p. 744. As adversaries of epigenesis, T. Huxley rightly cites the partisans of transformism, or, as Bonnet called it, evolution: Malpighi, who drew along in his train Liebniz and Malebranche. In sum, the doctrines first designated by the words "evolution" and "development" have been abandoned, but the words have survived and are applied to "every series of genetic changes observable in living beings." (p. 746)

6. In his notebook, under the date of July 1837, Darwin wrote: "In July opened first notebook on Transmutation of Species"; cited by Gertrude Himmelfarb, *Darwin and the Darwinian Revolution* [Gilson says *"Evolution"*] (New York: Doubleday, 1959), p. 146.

7. "The theory of modification through natural selection," *Origin*, ch. VI, p. 86 (p. 134). Cf. in the single final chapter of the work (ch. XV): "the theory of descent with subsequent modification," p. 233 (p. 358); "The theory of natural selection," p. 234 (p. 360); "the principle of natural selection," p. 237 (p. 364); "their (the species') long course of descent and modification," p. 237 (p. 364); "the mutation of species," p. 240 (p. 368); the last paragraph of the book, p. 243 (p. 374), does not speak of evolution, but the last word is "evolved." This is the closest that he may have come to the word using his

own language in his own name. The word already ended the *Sketch of 1842* (ed. Gavin de Beer, p. 87) and the *Essay of 1844* (ed. cit., p. 254). It has always been there.

8. *Nouveau petit Larousse illustré* (Paris, 1952), art. "Evolution": "Darwin upheld the doctrine of evolution." The Great Books, Syntopicon, ch. "Evolution," Introduction: "This chapter belongs to Darwin," vol. II, p. 451. The author of this article had indeed seen that Darwin does not ordinarily use the word, but that does not prevent him from concluding that it is to him that it properly belongs. Note the same remark in connection with John Ramsbottom, "Lamarck and Darwin," in *Précurseurs et fondateurs de l'évolutionnisme: Buffon-Lamarck-Darwin* (Paris: Muséum d'Histoire Naturelle, 1963), p. 25: "It is appropriate, however, to note that Darwin did not speak of evolution, but of descent [*descendance*] with modification." We shall see further on some reasons for this. Loren Eiseley, *Darwin's Century: Evolution and the Men Who Discovered It* (New York: Doubleday Anchor Books, 1958): This century, which is that of Darwin because he is one of the discoverers of evolution, appears to have heard little spoken of it; the "Index" to this book mentions the word "evolution" only one time: "human evolution." [This is simply not true: see "Index," p. 365. There are nine topical entries here under "Evolution" or "Evolutionary Theory," and thirty references to pages where the topics are discussed.] In fact, one could write a book closely following Darwin and never have the occasion to use the word. Benjamin Farrington, *What Darwin Really Said* (New York: Schocken Books, 1966), does not mention the fact that Darwin did not say "evolution"; on the contrary, see p. 117.

9. "The words 'evolve' and 'evolution' do not actually appear in Darwin's early writings, including the first five editions of the *Origin*. Although Lyell had used 'evolution' in its present sense in his *Principles*, and Spencer more prominently in his essay on 'The Development Hypothesis' in 1852, it was not then commonly used, and entered the popular and scientific vocabularies only later. 'Change', 'variation', 'transformation', 'transmutation', and 'mutability' were the accepted terms of the doctrine, with *'chain of being', 'tree of life',* or *'organization of life'* to connote the evolutionary hierarchy. *'Evolve'* and *'evolution'* appear in this discussion only when they convey the same meaning as Darwin's less familiar terms." Gertrude Himmelfarb, op. cit., notes to ch. 7, note 1, p. 442 [1962 ed., p. 467].

10. C.D. to A.R.W., May 1, 1857. In Francis Darwin, ed., *The Autobiography of Charles Darwin and Selected Letters* (New York: Dover, 1958), p. 193.

11. Ibid.

12. C.D. to J.D. Hooker, June 29, 1858. In F. Darwin, ed., op. cit., p. 198.

13. On the delicate problem the comparison of the two reports presents, see the quite attentive work of Georges Canguilhem, *Etudes d'histoire et de philosophie des sciences* (Paris: Librairie Philosophique J. Vrin, 1968), pp. 105-10. M.G. Canguilhem summarizes his thought on the question in this fashion: "What is to be concluded from this confrontation? This, namely that

if Darwin found in the writing of Wallace the essence of his own ideas in spite of the absence of the words "natural selection," it is because these words already designated in his own thought nothing other than the sum total [*totalisation*] of certain conceptual elements." (p. 107)

14. *Autobiography,* ed. F. Darwin, pp. 18-19.

15. Paley's book had been directed against antifixist doctrines such as those of Charles Darwin's grandfather, Erasmus Darwin, and Lamarck. "They would persuade us to believe that the eye, the animal to which it belongs, every other animal, every plant, indeed every organized body which we see, are only so many out of the possible varieties and combinations of being, which the lapse of infinite ages has brought into existence; that the present world is the relict of that variety; millions of other bodily forms and other species have perished, being by the defect of their constitution incapable of preservation, or of continuance by generation. Now there is no foundation whatever for this conjecture in anything which we observe in the works of nature; no such experiments are going on at present; no such energy operates, as that which is here supposed, and which should be constantly pushing into existence new varieties of being." William Paley, *Natural Theology*, 18th ed., 1818, cited by Benjamin Farrington, *What Darwin Really Said* (New York: Schocken Books, 1966), pp. 39-40. See pp. 41-42 for a critique of creationism implied in the teleology of Paley which could have played a determining role in Darwin's thought: The teleology of Paley had been conceived as necessarily created.

16. Genesis says in many places (1:12, 20, 24) that God created plants and animals (already bearing seed and capable of reproduction): "according to their kind" [*selon leurs espèces*]. Even the literalist which Darwin was at first could not read there that God had created each species by a distinct act, nor, still less, that he had created species fixed and such then as they are yet today. As much of a controversialist as Darwin was not, and much more contorted (he is known to have made inquiries about Suarez so that he might triumph over an imprudent theologian), Thomas H. Huxley appears to have perceived the weakness of the position. He thus surreptitiously displaced the ground of the argument from Creation to the Deluge. In his most remarkable article in the *Encyclopaedia Britannica* (9th ed. New York, 1878), vol. 8, p. 751, arguing from geographical distribution of species (ornithorhynchus [platypus] limited to Australia, and various sparrows to South America), Huxley concluded that when these facts were known "all serious belief of the peopling of the world from Mount Ararat would end." Thomas Huxley pretends to identify himself with the "uncultivated people" to whom the message of Moses was addressed. This is too modest. He would do better to scrutinize the mystery of what the deluge could have done to fishes.

17. "An Historical Sketch of the Progress of Opinion on the Origin of Species, Previously to the Publication of the First Edition of This Work [Gilson has "Book"]," ed cit., p. 1 [See *Origin*, p. 1 (p. 3).].

The *Sketch of 1842* and the *Essay of 1844* do not allow anyone to doubt that Darwin had seen the idea that each individual organism necessitated the act of a creator ("must require the fiat of a creator"), *Evolution by Natural Selection*, ed. Gavin de Beer, p. 87. Darwin appears to think here of the text of Genesis 2:19: "And out of the ground the Lord God formed every beast of the field and every fowl of the air, and brought them into Adam to see what he would call them." Darwin says further: "It is to derogate from the creator of innumerable universes, to believe that he should have made by individual acts of his willl, the myriads of creeping parasites and worms which, since the earliest dawn of life, have pullulated upon the land and in the depth of the Ocean." Op. cit., p. 253. But he can also speak exclusively of a creation of species. He appears then to remind himself of the aphorism of Linnaeus: "I repeat, ought we then say that a pair, or a pregnant [*pleine*] female, of each of the three species of rhinoceros, had been created separately with the false appearance of [having had] a true parent?" pp. 250-51. He does not appear to have taken the trouble to define in detail his adversary. What is important to him is to maintain that "the specific forms are not immutable creations." Op. cit., p. 252. The theological naivete of Darwin brings again to mind the fact that his provisional clerical vocation was above all the result of the failure of his medical vocation. With respect to the philosophers he himself said, in his *Autobiography*, that he knew little of them. He had been able to apprehend from Malebranche that "God never acts through particular acts of the will," but this *gentleman* [Darwin] was not a professional in anything, not even in the natural sciences. Only his own researches and his own ideas interested him.

It could easily be conceived that Darwin had been profoundly troubled by Wallace's memoir, which, on the basis of information quite slender when compared to his own, and in a style indulgent of generalities, affirmed: 1.) that the life of wild animals is a struggle for existence; 2.) that new species are formed by reason of the survival of individuals who bear variations which favor their survival. To compensate for not arguing like Darwin from domestication to natural selection, Wallace opposed the two modes of propagating species, domestic species, according to him, having a tendency to revert to their native parent stock if let alone, while wild species on the contrary have a tendency to form incessantly new varieties.

18. "An Historical Sketch," in *The Origin of Species*, ed. cit., p. 1 [*Origin*, p. 1 (pp. 3-4)]. Darwin gave a place to Lamarck from the first redaction of the *Origin*, ch. 1: "Something may be attributed to the direct action of the conditions of life." In the fifth edition, 1869, Darwin corrected this to read: "to the definite action of the conditions of life, but how much we do not know." He preferred "definite" to "direct"; it is less definite. A second improvement, in the sixth edition (1872): "something, but how much we do not know" etc. (*A Variorum Text*, p. 118). The phrase that follows ("something must be attributed to use and disuse") is likewise corrected in the sixth edition, 1872: "Some, perhaps a great, effect may be attributed to the increased use or disuse of parts." This second element of Lamarckism appeared to him to

deserve a bit more attention than the first, but he is not ever much interested in it either; which does not prevent him from often falling into Lamarckian arguments in the course of his own explications.

19. "An Historical Sketch," ed. cit., p. 4 [*Origin*, p. 4 (p. 8). Gilson has "1958" for "1858."].

20. Darwin, *Origin*, ch. XV. "Recapitulation and Conclusion," pp. 240-41 (p. 369).

21. Darwin, *The Descent of Man*, ch. II, p. 285.

22. *The Autobiography of Charles Darwin*, ed. cit., p. 260. [Italics are in the English edition.]

3. Evolution without Darwin

1. *Origin*, ch. XV, "Conclusion," p. 240 [p. 369].

2. Herbert Spencer, *Le principe de l'évolution*, ed. cit., p. 4. I have not seen the German translation. The French translation differs from the original in a rather important respect. It will be noted in due time. The French brochure is an extract [*tiré à part*] from the *Journal des Economistes* (Dec. 15, 1895), pp. 740-57. The French version is preceded by an introduction in which Spencer explains why he is responding. It is that the members of the French Academy of Sciences have approved the presentation made to that body of a French translation of Lord Salisbury's discourse (*Journal des Economistes*, art. cit., p. 320). This event, reported by English journals with favorable commentaries, had convinced Spencer of the opportunity to react so that public opinion would not be misled on the question. The English original is found under the title, (Herbert Spencer) "Lord Salisbury on Evolution, Inaugural Address to the British Association, 1894," in *The Nineteenth Century*, November 1895. [All further references to this essay will be to the English edition and its pagination].

3. [That has not been obvious to commentators. See, for instance, Charles Saunders Peirce's observations about the noninductive nature of Darwin's hypothesis. *Collected Papers*, 6, 297, cited in Philip R. Wiener, *Evolution and the Founders of Pragmatism* (New York: Harper, 1965), p. 78. For Darwin's vacillation between the primacy of "theory" or that of fact-finding, see Nora Barlow's "Appendix" to her edition of *The Autobiography of Charles Darwin* (New York: Norton, 1969), part I, "On Charles Darwin and His Grandfather Dr. Erasmus Darwin," pp. 157-66. See also Darwin's letter to Henry Fawcett (Sept. 18, 1861): "You will have done good service in calling the attention of scientific men to means and laws of philosophizing. . . . How profoundly ignorant B. must be of the very soul of observation! About thirty years ago there was much talk that geologists ought only to observe and not theorise; and I well remember some one saying that at this rate a man might as well go into a gravel-pit and count the pebbles and describe the colours. How odd it is that anyone should not see that all observation must be for or against some view if it is to be of any service!" Letter to Henry Fawcett, in

More Letters of Charles Darwin, ed. Francis Darwin and A. C. Seward (London: John Murray, Albemarle Street, 1903), vol. I, p. 195. Darwin was a cryptometaphysician, or at least philosopher, despite his periodic disclaimers to the contrary, and Gilson is quite aware of this. See, e.g., "Appendix II," below, p. 252.]

4. [Herbert Spencer, "Lord Salisbury on Evolution," *The Nineteenth Century* (November 1895), p. 740.]

5. [Spencer, loc. cit., pp. 740-41. Gilson apparently uses the French translation, which is only slightly different here.]

6. [I am translating from Gilson's French here, not being able to find the passage in the English original. There is something like it on p. 105, loc. cit., however.]

7. Spencer, loc. cit., p. 741.

8. [Spencer, quoting from an essay of his from 1852, loc. cit., pp. 741-42.]

9. [Ibid., p. 742.]

10. For the rejoinder to creationism which he judges invincible, see, for example: "Nobody has seen a species evolved and nobody has seen a species created." [p. 743] True. But this is why evolution ought not be presented as a scientific truth, whereas it is permissible to believe in creationism without scientific proof, as a metaphysical or religious truth. The two cases are not of the same order.

11. [Spencer, loc. cit., p. 742.] It is beyond doubt that Spencer was at the origin of the movement which made of the notion of evolution the key word in the thought of the years 1850-1910. The fusion of Darwinism and Spencerism was almost instantaneous, as it seems, in spite of the ill will of the authors. We only note the fact here, still on the terrain of biology where the problem of teleology is presented. I dare not say if the psychological and ethical aspect of Darwin's thought did or did not contribute to this fusion through joining together again the moral and social speculations of Spencer to constitute "Social Darwinism," which was so lively in the United States. It appears highly probable, but the history of the fusion needs to be written. We content ourselves with observing that at the very time Darwin and Spencer lived the thing was already done, and, it appears, on the ground of biology alone. I have not myself encountered social Darwinism in the preceding research, but this research is limited, and I dare affirm nothing beyond that.

12. [Spencer, loc. cit., p. 742.] Spencer was naturally far from holding the two theories as rational equivalents. Suppose that there are, or that there have been, ten million species: "which is the most natural theory about these ten millions of species? Is it most likely that there have been ten millions of special creations (each implying a conscious design and acts in pursuance of it)? or is it more likely that, by continual modification due to change of circumstances, ten millions of varieties (i.e., kinds) have been produced? (sic)" [Spencer, citing his 1852 work, loc. cit., p. 742.] How have these creations taken place? "If they have formed a definite conception of the process, let them tell us how a new species is constructed, and how it makes its appear-

ance. Is it thrown down from the clouds? or must we hold to the notion that it struggles up out of the ground? Do its limbs and viscera rush together from all the points of the compass? or must we receive the old Hebrew idea, that God takes clay and molds a new creature? (sic)" [ibid.] For himself, in the last analysis, Spencer concludes on the basis [*foi*] of indirect evidence that "the idea of a special creation, when brought distinctly before us by alleged cases, is too absurd to be entertained." [Spencer, 1895, p. 747. Gilson cites, from the French, the same page as the previous two quotations.]

13. "Naturalists continually refer to external conditions, such as climate, food, etc., as the only possible cause of variation. In one limited sense, as we shall hereafter see, this may be true; but it is preposterous to attribute to mere external conditions, the structure, for instance, of the woodpecker, with its feet, tail, beak, and tongue, so admirably adapted to catch insects under the bark of trees. In the case of the mistletoe, which draws its nourishment from certain trees, which has seeds that must be transported by certain birds, and which has flowers with separate sexes absolutely requiring the agency of certain insects to bring pollen from one flower to the other, it is equally preposterous to account for the structure of this parasite, with its relations to several distinct organic beings, by the effects of external conditions or of habit, or of the volition of the plant itself." Darwin, *Origin,* "Introduction," pp. 6-7. The last trait marks a state already advanced beyond the critique of Lamarck. He has been reproached with allowing [*admettre*] in living being a *will* to adapt. Darwin does not appear to have perceived that this critique was the exact match of that which was addressed to him, when he was reproached, concerning natural selection, with attributing to nature a faculty of exercising a "choice."

14. Darwin, *Origin*, ch. IV, p. 40. [Italics and English text given by Gilson.]

15. Spencer, loc. cit., p. 757.

16. Spencer, *First Principles* [Osnabruck: Otto Zeller, 1966. Reprint of 1904 ed.], §185 [p. 433].

17. [*Origin*, p. 4 (p. 8).]

18. [Ibid. The passage cited follows after the one cited in the previous footnote, with no intermediate reference to Lamarck.]

19. Op. cit., chs. XII-XVIII.

20. Spencer, *Les premiers principes*, trans. E. Cazelles (Paris: Germer-Baillière, 1871), p. 326. [Zeller reprint, §105, p. 244. This seems to be the passage Gilson is citing.]

21. Spencer, *Les premiers principes*, ch. XIV, §110, p. 332; §105, pp. 325 and 326. [See Herbert Spencer, *First Principles*, Zeller edition, §110, p. 249; §105, pp. 243, 244. §105, however, is in ch. XIII, and not in ch. XIV as Gilson states.]

22. *The Autobiography of Charles Darwin, 1809-1882*, ed. Nora Barlow (London: Collins, 1958), pp. 108-9.

23. G. Himmelfarb's work contains a portrait of Spencer painted in lively colors, without undue indulgence; in sum, true. Op. cit., pp. 213-14.

24. [Spencer cites only essays of 1857 in this "Preface."]

25. [Gilson has "xxx."]

26. H. Spencer, "Preface to the Fourth Edition," of *First Principles* (May 1880), op. cit., p. viii. Darwin is cited four times in the *First Principles:* §133, §159 (important), §166 (on the divergence of characters) [sic–only three times listed. §133 is not cited in the "Index" to this edition–but the reference to Darwin is there. Again, the "Index" cites §110 (p. 253), while Gilson does not notice it.]. In the article in the *Westminster Review*, April 1, 1857, after having formulated what he calls then "the law of organic progress," as if progress and evolution were equivalent terms, Spencer already announced his intention of extending his law to the history of the earth, to life, society, government, commerce, language, literature, art, in brief, to everything. If one takes the notion of evolution with this degree of universality, where it rejoins Heraclitianism, it is to be found everywhere before Spencer. When one reads the eminently readable book of Loren Eiseley (*Darwin's Century: Evolution and the Men Who Discovered It*), it appears that many people may have discovered it, even Linnaeus, the patriarch of fixism. Indeed, almost everyone has a claim to it, except Spencer, who gets only two lines and one note (pp. 215-16, and p. 313, n.): "Herbert Spencer, one of the English pre-Darwinian evolutionists." One is tempted to think that this historian speaks of Spencer without being aware of how appropriately he speaks. One could only with difficulty find a better proof of the elimination of the theoretician of evolution by Darwin, who is hardly interested in theory. To judge of it on the criterion of evolution, the ninteenth century ought rather be called "The Century of Spencer." No one thinks of it as such.

27. [Spencer, *First Principles*, "Preface to the Fourth Edition," p. viii. Not cited by Gilson.]

28. ["Ce nouvel *hircocervus*." Actually, the *hircocervus* was a mythical animal, part goat, part deer.]

29. On an important point Spencer obtained from Darwin a change of vocabulary. He made the observation to Darwin that the expression "natural selection" was equivocal: it invited the personalization of nature, the imagining of her choosing after the fashion of a stockbreeder who proceeds consciously to his choices. Spencer proposed in its place "survival of the fittest." Darwin largely deferred to his suggestion, while observing, moreover, that the expression "natural selection" was a metaphor whose meaning could hardly deceive anyone. See L. Eiseley, op. cit., p. 748 [the page number is certainly wrong, for there are only 378 pages in the volume. Nor is it obvious to what page or pages Gilson might be referring. This may be a printing error in the French edition.].

30. Francis Darwin, *The Autobiography of Charles Darwin and Letters* (New York: Dover, n.d.), ch. IX, p. 175.

31. We regard Spencer here, from the outside, as a philosopher. He himself had been convinced that his theory of evolution rested on solidly scientific bases, which, moreover, he did not claim to have discovered. In his response to Lord Salisbury he cites four great groups of facts as telling the same story: fossils, the truths of classification, the distribution of species in space, and embryology ("Lord Salisbury on Evolution" [pp. 743-45 (Gilson cites p. 745)]). The French text, instead of having "these four great groups of facts" has "now, from these five orders of facts" and adds a paragraph which develops the brief suggestions of the English text concerning rudimentary organs which are quite sensible in light of the hypothesis of evolution, and quite nonsensical upon the contrary supposition. ["Lord Salisbury on Evolution," pp. 745-46.] The French addition goes from "To the facts drawn from embryogeny" to "of maladies often mortal."

32. G. Himmelfarb, op. cit., ch. XV, pp. 297-98 [1962, p. 312].

33. "Lord Salisbury on Evolution," pp. 745-46.

34. [Thomas Henry Huxley, "Evolution" and "Evolution in Biology," in Encyclopaedia Britannica, 9th ed. (New York: Samuel Hall, 1878), vol. VIII, p. 749. The article on "Evolution" runs from p. 744 to p. 773. No citation is given by Gilson.]

35. [Sully, "Evolution in Philosophy," loc. cit., p. 746.]

36. [Ibid.]

4. Darwin and Malthus

1. "In October 1838, that is, fifteen months after I had begun my systematic inquiry, I happened to read for amusement Malthus on *Population*, and being well prepared to appreciate the struggle for existence which everywhere goes on from long-continued observation of the habits of animals and plants, it at once struck me that under these circumstances favourable variations would tend to be preserved and unfavourable ones to be destroyed. The result of this would be the formation of new species. Here, then, I had at last got a theory by which to work, but I was so anxious to avoid prejudice, that I determined not for some time to write even the briefest sketch of it. In January 1842 I first allowed myself the satisfaction of writing a very brief abstract of my theory in pencil in 35 pages; and this was enlarged during the summer of 1844 into one of 230 pages, which I had fairly copied out and still possess." *The Autobiography of Charles Darwin and Selected Letters*, ed. Francis Darwin (New York: Dover, 1958 [Gilson says, *"sans date"*]), pp. 42-43.

For those who know the complete candor of Darwin, this testimony is literally true. He did not say, there or elsewhere, that he owed the idea of natural selection to Malthus. On the contrary, this passage follows another in which he expressly says: "I soon perceived that selection was the keystone of man's success in making useful races of animals and plants. But how selection could be applied to organisms living in a state of nature remained for some time a mystery to me." Ibid., p. 42. In addition, although he might have been able to find in Malthus direct applications of the law of population to plants and animals, Darwin did not appear to be conscious of owing anything to Malthus on this point. It is *his* experience as a naturalist, incomparably richer than that of Malthus, which was cleared up by Malthus' book. There he found the reason for natural selection, that is, the permanent and necessary disproportion between the increase of the means of nourishment and that of population.

One can consult with advantage the critical discussion on this point by Camille Limoges, *La sélection naturelle* (Paris: P.U.F., 1970), pp. 28-31, 77-81. It is difficult to weaken the decisive testimony of Darwin himself in the "Introduction" to the *Origin*: "I will then pass on to the variability of species in a state of nature. . . . We shall, however, be enabled to discuss what circumstances are most favourable to variation. . . . The struggle for existence amongst all organic beings throughout the world, which inevitably follows from the high geometrical ratio of their increase, will be considered. This is the doctrine of Malthus, applied to the whole animal and vegetable kingdoms." [*Origin*, p. 7 (pp. 12-13). Gilson does not give the precise citation, nor does he use elision marks for material which he leaves out of the quotation.] *This is the doctrine of Malthus*, which alone explains how the fittest are able to have a better chance of survival, "and thus be naturally selected." [Ibid., p. 7 (p. 13).] Malthus thus put Darwin on the way to the solution of a problem which he himself had not set.

The exact place of Malthus' principle is marked again with exactitude in the last sentence of the *Origin*: "a ratio of increase so high as to lead to a struggle for life, and as a *consequence to natural selection*, entailing divergence of character and the extinction of less-improved forms." [Ibid., p. 243 (pp. 373-74). Italics are Gilson's.]

2. This is not a question of an analogy of situations. We have noted the importance which the problem of scientific truth/revealed truth had in Darwin's eyes. A note of C. Limoges (*La sélection naturelle*, p. 152) puts us on notice that the fact has been already emphasized: "In sum, W. F. Cannon ("The Bases of Darwin's Achievement: A Reevaluation," in *Victorian Studies* [1961-1962]: 109-34), may have had reason to insist on the importance of natural theology in the birth of Darwinism. But what that theology furnished was not the framework of the new theory, but rather *the ground of the breakup*." If W. P. Cannon really spoke of the conflict with "English natural theology," it would suitably be rectified by saying simply "with theology," for the crisis of which Darwin himself spoke many times takes place on the ground of faith in the account of Genesis, which like most of his contemporaries (but not all) he judged irreconcilable with the transformation of species.

3. "Heaven forfend me from Lamarck nonsense [sic–English] of a 'tendency to progression', 'adaptations from the slow willing of animals', etc. But the conclusions I am led to are not widely different from his; though the means of change are wholly so. I think I have found out (here's presumption!)

the simple way by which species become exquisitely adapted to various ends." Letter to Hooker, in *Autobiography*, ed. Francis Darwin, p. 184. This text is inexhaustible: 1) no "tendency to progress," which distinguishes Darwin from the progressivist lineage of Lamarck, Spencer, Bergson, etc.; 2) the same [sort of] misinterpretation by Darwin about "willing" in Lamarck that Darwin justly reproaches others with having committed concerning "selection" in his own doctrine; 3) the novelty of his doctrine is not in the mutability of species but in the explanation of how they change; 4) the refinement of the adaptation of species and their variations to their ends.

4. To say that the analogy between natural selection and artificial selection, or domestication, only occupies a secondary place in Darwinism would be to go against the reiterated declaration of Darwin himself. He always held that idea as one of the most fecund that he had, and he attributed the cause of the errors of some others to the fact that they did not have it. "With respect to books on this subject (the mutability of species), I do not know of any systematical ones, except Lamarck's which is veritable rubbish; but there are plenty, as Lyell, Pritchard, etc., on the view of the immutability. Agassiz lately has brought the strongest argument in favor of immutability. Isidore G. St. Hilaire has written some good essays, tending towards the mutability side, in the *Suites à Buffon*, entitled *Zoologie générale*. I believe all these absurd views arise from no one having, as far as I know, approached the subject on the side of variation under domestication, and having studied all that is known about domestication." Letter to J. Hooker (1844), in *Autobiography*, ed. F. Darwin, p. 184. For the remarks which follow, see "Appendix II," on artificial selection as if [*quasi*] "unconscious."

5. T. Malthus, *On the Principle of Population, As It Affects the Future Improvement of Society* (London, 1836), vol. 1, pp. 6 and 517 (sic). Darwin cited Malthus then after the sixth edition of the work, itself a reimpression of the fifth edition, revised, published in 1817. The reference from the *Descent of Man* to Malthus is found in part I, ch. II, p. 175 (ed. cit.). On the problem of the Darwin-Malthus relationship see C. Limoges, op. cit., pp. 77-81. Malthus rendered the image of the struggle for life vivid and gripping for him. He gave him an intellectual shock. Darwin's notebook even speaks of *one* phrase of Malthus as its cause: "Population increases geometrically in a time *VERY MUCH LESS* than 25 years; however, until Malthus phrased it, no one clearly perceived the great obstacle which retarded it in mankind." C. Limoges, op. cit., p. 78, note 3. To which he adds: "This passage of Malthus' has been identified by Sir Gavin de Beer in the sixth edition of the *Essay*, I, p. 6: 'It may safely be pronounced, therefore, that the population, when unchecked, goes on doubling itself every twenty-five years, or increase [sic: Gilson] in a geometrical ratio'." Darwin had already come across a similar notion in de Candolle ("all the plants of a land, all those of a given place, are in a state of war with each other" etc., text in C. Limoges, op. cit., p. 65), but, for whatever reason it might be, Darwin says that it is Malthus' passage which struck him. Perhaps his mind was not ripe for the message when he read de Candolle, or

perhaps simply the message struck him more directly in English than in French, with which, as with German, he was never at ease. To speak truly, one does not know.

6. Cited by G. Himmelfarb in the "Introduction" to her edition of T. Malthus, *On Population* (New York: Random House-Modern Library, 1960), p. xxvi.

7. T. Malthus, *An Essay on the Principle of Population*, ed. G. Himmelfarb; these texts are found in what is ordinarily called *The First Essay* (1798), ed. cit., ch. I, pp. 8-9, and ch. II, p. 11. On the *superfecundity* of nature, and how it exceeds the means of sustenance for the beings which nature engenders, see William Paley, *Natural Theology* (London, 1821), ch. XXVI, pp. 394-95.

8. [That is hardly "certain." The contexts in which Malthus includes *vice* as well as misery as one of the generic means of limiting population allow one to infer that he would include at least *some* forms of contraception (to say nothing of abortion, etc.) as *vicious*. Gilson's statement in the next paragraph that, according to Malthus, nature alone should be entrusted with punishing the poor man who improvidently marries hardly bears out his suggestion of Malthus' presumed liking for legal measures for the limitation of birth were the parson alive today. For "vice" and "misery" see Thomas Malthus, *An Essay on the Principle of Population* (Harmondsworth: Penguin Books, 1970), pp. 71-72, 85, 89, 102-3, 139.]

9. T. Malthus, *An Essay on the Principle of Population or, a View of Its Past and Present Effects on Human Happiness* (sometimes called "the second essay"), ed. cit., pp. 530-33.

10. T. Malthus, op. cit., "First Essay," ch. I, ed. cit., pp. 9-10. [See T. Malthus, *An Essay on the Principle of Population*, ed. Anthony Flew (Harmondsworth: Penguin Books, 1970), pp. 71-72.]

11. I say "almost the only," because there was at least one other who found what he needed, and that one, by an almost unbelievable coincidence, turned out to be Wallace. See his letter to A. Newton, December 3, 1887: "But at that time I had not the remotest notion that he had already arrived at a definite theory – still less that it was the same as occurred to me, suddenly, in Ternate in 1858. The most interesting coincidence in the matter, I think, is, that I, *as well as Darwin*, was led to the theory itself through Malthus – in my case it was his elaborate account of the action of 'preventive checks' in keeping down the population of savage races to a tolerably fixed but scanty number. This had strongly impressed me, and it suddenly flashed upon me that all animals are necessarily thus kept down – 'The struggle for existence' – while *variations*, on which I was always thinking, must necessarily often be *beneficial*, and would then cause those varieties to increase while the injurious variations diminished." *Autobiography*, ed. cit., pp. 200-1. The whole history [of this discovery] abounds in details well set to disconcert the historian, who is always more the friend of the plausible [*vraisemblable*] than the true. It is, in any case, disarming.

12. We find ourselves to be in substantial accord with the conclusion of Camille Limoges: "That which Malthus would have supplied to Darwin is not the idea of the struggle for existence, then so common. Rather it would have been the idea of the intensity of that struggle, of its restraining power on living beings, the idea of a geometric progression implying that a constant 'pressure' is exercised on beings [*vivants*], necessarily engendering an incessant warfare among them, an ancestral form of the *population pressure* of present day population genetics." *La sélection naturelle*, p. 79. The author immediately adds: "That, and nothing more." This restriction signifies that, according to him, "it is doubtful that this contribution of Malthus had been strictly [*de droit*] indispensable to the constitution of the theory." (P. 79; cf. p. 152.) There would be then two histories of science, that which consists essentially in the intellectual biography of scientists and that which only seeks to comprehend "the formation and the transformations of concepts, scientific theories, and methods of research." The problem is presented in analogous terms for the history of philosophy, where one observes often also impersonal necessities of thought and biographical contingencies; but, in the end, philosophy only exists by virtue of philosophers, as science by virtue of scientists; the contingency inherent in the order of human events, even if one wishes to remove it from science, is at least inseparable from its history.

13. "Struggle for Existence," Malthus, "First Essay," ch. III.

5. *Evolution and Teleology*

1. Darwin to J. D. Hooker [Gilson has "Hosker"], September 25, 1853. In *Autobiography*, ed. F. Darwin, p. 188. It is true that Darwin, who knew himself well, adds immediately: "But I must confess that perhaps nearly the same thing would have happened to me on any scheme of work."

2. Samuel Buker, cited by G. Himmelfarb, op. cit., ch. XV, p. 305.

3. Darwin thought about the question nevertheless, but knowing himself incapable of answering it, he avoided posing it. Here, from another source, is what he says of it: "Looking to the first dawn of life, when all organic beings, as we may believe, presented the simplest structure, how, it has been asked, could the first steps in the advancement or differentiation of parts have arisen? Mr. Herbert Spencer would probably answer that, as soon as simple unicellular organism (sic) came by growth or division to be compounded of several cells, or became attached to any supporting surface, his law 'that homologous units of any order become differentiated in proportion as their relations to incident forces become different' would come into action. But as we have no facts to guide us, speculation on the subject is almost useless. It is however an error to suppose that there would be no struggle for existence, and, consequently, no natural selection, until many forms had been produced. . . . But, as I remarked towards the close of the Introduction, no one ought to feel surprise at much remaining as yet unexplained on the origin of species, if we make due allowance for our profound ignorance on the

mutual relations of the inhabitants of the world at the present time, and still more so during past ages." [*Origin*, ch. IV, Great Books ed., p. 62 (M.L. ed., p. 96)]. It appears that especially since it is impossible to define scientifically a species, one ought to cease to hold the concept as a scientific one, and content oneself to use it empirically according to commonsense, in which it suffices to distinguish species for stockbreeding and zoological gardens. It is not reasonable to search for the origin of an object of observation which it is recognized is incapable of definition.

4. Darwin, *Origin*, ch. VI [Gilson says ch. V. Great Books ed., pp. 94, 95 (M.L. ed., pp. 146, 148)].

5. Darwin, *Autobiography* [ed. Francis Darwin: Letters to J. D. Hooker, July 12, 1860, and "1861"], pp. 322 and 324. [The second quotation above occurs first in the letters.]

6. L. Cuénot's book *L'adaptation* (Paris: G. Doin, 1925, Bibliothèque de Biologie générale, dir. M. Caullery) sets forth quite useful scientific distinctions between accommodation (individual adaptation), acclimatization (of a species which only thrives by virtue of man's care), and naturalization (or specific adaptation), when the species becomes a permanent part of its new milieu. "An adaptation is in reality the solution of a problem, exactly as a machine or an utensil made by man." Adaptation here is only another name for finality. The fourth part of the work ("The Metaphysics of Adaptation") subscribes (p. 389) to the conclusion of Ch. Richet (1913): "If life has emerged from inert matter, if intelligence has struggled up out of the unconscious, it is because a law directed the cosmic forces to that end [*dans ce sense-là*]. No one dare say that this law has *willed* life and intelligence, for the verb 'to will' is terribly human. But no one can refuse to recognize that the gradual development of life and intelligence was a foreordained part of the terrestrial globe" (pp. 389-90). One recognizes here the venerable anthropocentrism of the Bible, transformed, with the verbosity of Teilhard de Chardin. On this notion in Darwin himself see Camille Limoges, "Darwinisme et Adaptation," in *Revue des Questions scientifiques*, XIV, no. 3 (July 1970): 353-74. This study considers the notion of adaptations as an element [*une pièce*] not necessary to Darwin's doctrine. Darwin appears to have used the word in the usual sense, without ever distinctly troubling himself about its abstract sense. "Adaptation" signifies for him adaptations, the things adapted to each other. Unless it be an error on my part, he continually spoke of it in the *Origin* without ever having "posed the problem of adaptation" there. In the usual sense the word can be defined, in the passive voice, as the adjustment of two or more things to a common situation or a common function; or, in the active voice, as "the process by which one thing is modified in order to be able to enter into a new adjustment." (C. Limoges) But a *process in order to be able* is not essentially distinguished from final causation [*d'un rapport de finalité*]. If the word "adaptation" only signifies the brute fact, it poses no philosophical problem, and one indeed has the impression that Darwin did not use it otherwise. If it signifies an abstract notion, philosophical or scientific, it presents

a problem, and not a solution. Adaptation is not a simply epistemological obstacle; it is an obstacle in reality. One says "adaptation" in order to avoid saying "finality." On the unavoidability of this, "that the teleological problem cannot not be posed," and that "we run into it as much by the *how* as by the *why*," see the useful dialectic exercise proposed by Eugène Ionesco, *Présent passé Passé présent* (Paris: Mercure de France, 1968), pp. 136-37.

7. *Autobiography* [ed. F. Darwin], p. 308.

8. [Ibid., p. 316.]

9. The combination of Darwin and evolution, generally prevalent, is, however, established with particular force in the United States, where Darwin has become the prophet of a rationalist, antireligious reaction, or at least one that is antibiblical. This was not to travesty his thought, for he himself had to choose between natural selection and that which he thought was the teaching of Scripture. The history of Darwin in the United States is a peculiar chapter in that of evolutionism. On this subject consult: G. Daniels, *Darwinism Comes to America,* Blandel paperback [sic]. Under the same title, R. J. Loewenberg, published by R. C. Wolf, Fortress paperback [sic].

10. Thomas Huxley's text cited by Francis Darwin appears in an essay on the "Geneology of Animals," in *The Academy*, 1869, reprinted by Huxley in his *Critiques and Addresses*, p. 305, and, finally, cited by Francis Darwin, *The Autobiography*, p. 316. The same idea has been taken up by many successors of Darwin in their own accounts. In this connection see L. Cuénot, *Invention et finalité en biologie*, pp. 94-95, notably the passage taken from De Vries: "The high value of Darwin's theory of selection, as everyone recognizes, consists in explaining finality in organic nature by means of purely natural principles and without recourse to any theological idea." Darwin has also been praised for having pointed out a *teleonomy* which hides this finality so that one need not see it. The humorous formula of this operation is that of Jacques Monod: "Teleonomy is the word one ought to use if, by virtue of modesty, one prefers to avoid using *finality*." ("Leçon inaugurale de la chaire de la biologie moléculaire," November 3, 1967. Collège de France, no. 47, p. 9.) Speaking of the "fundamental property of all living beings without exception, that of being objects endowed with a project which at one and the same time they represent and accomplish by their operations," the same scientist declares: "rather than reject [*refuser*] this notion (as certain biologists are tempted to do), it is on the contrary necessary to recognize it as indispensable to the very definition of living beings. We shall say that these are distinguished from all other structures of all the systems present in the universe by that property which we call *teleonomy." Le hasard et la nécessité. Essai sur la philosophie naturelle de la biologie moderne* (Paris: Le Seuil, 1970), p. 22. On this surrogate for teleology see further pp. 27, 29, 32-33. "The organism is a machine which builds itself" (p. 60); it is a chemical machine constructed and held together by proteins, whose "teleonomic performances" belong finally to "their properties called *stereospecific,* that is to say, their capacity to

recognize other molecules (composed of other proteins) according to their *form* which is determined by their molecular structure. It is, literally, a question of a microscopic discriminative property (otherwise, 'cognitive')" (p. 60). A teleonomy immanent in life and analogous to consciousness differs from classical teleology in name only.

11. [Darwin to W. Graham, July 3, 1881. In *Autobiography*, ed. F. Darwin, p. 68. Not cited or completely dated by Gilson.]

12. The Duke of Argyll, *Good Words*, April 1885. Cited by F. Darwin, ed., *Autobiography*, p. 68.

13. One still finds biologists attached to the theory of evolution by way of natural selection. They seem to be more numerous among biochemists than among zoologists. François Jacob (*La logique du vivant.* Paris: Gallimard, 1970) is a good example of them. He seems to hold as established that under some form or another natural selection has been demonstrated by naturalists. Those who think thus do not themselves bring any demonstration of the fact. They do not even take the pains of refuting objections raised by other naturalists against the doctrine. Lemoine's thesis does not have any place in François Jacob's book, any more than does Vialleton's, some of whose arguments are more especially efficacious since they are developed on a strictly mechanistic plan: *Morphologie générale. Membres et ceintures des vertébrés tétrapodes. Critique morphologique du transformisme.* Paris, 1924. The more transformism is viewed at a distance, the less difficulty it presents.

14. To take exception to evolutionism is, in fact, to contest the possibility of the transformation of one species into another; it is not to subscribe to fixism. Species can disappear, others appear; and it is possible to have among them analogy without filiation. All filiation, if it takes place with them [*en*], remains within the species, as is the case with the group [*le groupe*] of *Equidae*. Species is without doubt a more supple and plastic concept than one imagines it to be: it is not a logical definition. The conclusions of Lemoine can be balanced by those of Etienne Wolf, *Les chemins de la vie* (Paris: Hermann, 1963), pp. 162-66.

15. [Gilson gives no citation for any of the quotations from Lemoine. The passages he cites, however, are scattered throughout pp. 5.82.3–5.82.8 (sic) of the section entitled "Que valent les théories de l'évolution?" in "Conclusions Générales," t. V, *Les êtres vivants*, of the *Encyclopédie française* (Paris: Société Nouvelle de l'Encyclopédie Française, Librairie Larousse, 1965). The entire section "Que valent les théories de l'évolution?" is one of Lemoine's contributions to the *Encyclopédie.*]

16. Jean Rostand, "Le problème de l'évolution," in *Les grands courants de la biologie* (Paris: Gallimard, 1951), p. 176.

17. Jean Rostand, *L'évolution des espèces* (Paris: Hachette, 1932), p. 191.

18. Ibid. Cf. *Les grands courants de la biologie*, p. 178. A new scientific defender of finality as a fact enters the lists: Pierre-P. Grosse of the Academy of Sciences: *Toi, ce petit Dieu! Essai sur l'histoire naturelle de l'homme* (Paris:

Albin Michel, 1971), especially pp. 46, 55-63. See p. 62: "Teleology in fact, such as we ascertain it in every living being, . . . is not a mental construction; it exists, and to deny it is to deny the biological fact itself."

Chapter IV. Bergsonism and Teleology

1. Paul Janet, *Les causes finales*, 2nd ed., rev. (Paris, 1882). The first edition dates from 1876. It is this one which Bergson cites in *Evolution créatrice* (in *Oeuvres*, ed. A. Robinet and H. Gouhier [Paris: P.U.F., 1959]) [Gilson has 1859], p. 547. This fact appears curious, for *Evolution créatrice* was published in 1907. The fourth edition of Janet's book, which was a reproduction of the second edition, having appeared in 1901, Bergson would have been able to cite an edition containing the very important preface written by Janet for the second edition. It could be supposed (gratuitously, moreover) that Bergson acquired at an early age a copy of the first edition (he was 17 years old in 1876) and neglected to consult a more recent edition; or, further, perhaps it was that the essence of classical finalism then appeared to him to have been fixed once for all.

2. Paul Janet, *Les causes finales*, ed. cit., "Preface."

3. Laplace, *Introduction à la théorie analytique des probabilités*, in *Oeuvres complètes* (Paris, 1886), vol. VII, p. vi; cited by Bergson, *Oeuvres*, pp. 526-27. [*Evolution créatrice* will hereafter be cited according to the English translation of Arthur Mitchell in the Modern Library edition: *Creative Evolution* (New York: Random House, 1944). For the citation from Laplace see pp. 43-44. For the passage itself see Pierre Simon, Marquis de Laplace, *A Philosophical Essay on Probabilities*, trans. F.W. Truscott and F.L. Emory (New York: Dover, 1952), p. 4.]

4. Cited by Bergson, *Oeuvres*, p. 527 [Mitchell, p. 44], without indication of origin. These examples are excellent, but in biology very few moderns are equal to the boldness of the mechanism of some of the ancients. See the critique of the mechanist biologies of Epicurus and Aesclapius by the peripatetic [philosopher] Galien (second century before Christ) [sic]; *Des facultés naturelles*, I. I, ch. 12, sv. [Gilson apparently means Galen, who lived during the second century *after* Christ, favored peripateticism, and composed a work known in English as *On the Uses of the Parts of the Body of Man.*]

5. *La pensée et le mouvant,* in *Oeuvres,* p. 1256.

6. Bergson did not hold evolution to be a strictly demonstrated truth, but he did hold it as scientifically certain in its order: "so that, all things considered, the transformist hypothesis looks more and more like a close approximation to the truth. It is not rigorously demonstrable, but failing the certainty of theoretical or experimental demonstration, there is a probability which is continually growing, due to evidence which, while coming short of direct proof, seems to point persistently in its direction: such is the kind of probabil-

ity that the theory of transformism offers." *L'évolution créatrice,* in *Oeuvres,* p. 515 [Mitchell, p. 29]. Even if transformism were proved false, a double thesis would remain nevertheless: 1) classification, which would remain in any case, presumes "this relation of, so to speak, *logical* affiliation between the forms"; 2) since the facts of paleontology would remain also, it still would be necessary to admit a *chronological* affiliation between the forms: "now, the evolutionist theory, so far as it has any importance for philosophy, requires no more." Ibid. [Mitchell, pp. 29-30.] For the connection with Spencer see "Le mécanisme et la vie," in *L'énergie spirituelle* [1919], ed. cit., p. 828, for purposes of comparison with the response of Spencer to Lord Salisbury already cited, see [pp. 120 and 130, above].

7. *L'évolution créatrice,* "Introduction," pp. 490-91 [Mitchell, pp. xx-xxi].

8. Ibid. [I do not find this passage anywhere in the English translation].

9. Ibid., p. 493 [Mitchell, p. xxiv].

10. Ibid., pp. 527-28 [Mitchell, p. 45].

11. Ibid., p. 528 [Mitchell, p. 45].

12. Ibid., p. 571 [Mitchell, p. 99].

13. Ibid.

14. Ibid.

15. Darwin, whom Bergson appears to consider as a mechanist in biology, quite strongly saw the point, on which Bergson ought to insist strongly, that the existence of homologous structures in divergent lines of evolution cannot be explained by "mechanical principles." According to Darwin "homologous structures are inexplicable by the simple principle of adaptation." To his liking, no one has shown as well as Professor Bianconi "how admirably such structures are adapted to their final design," but, he naturally adds, this adaptation can only be explained by natural selection: *The Descent of Man,* I, 1; Great Books, vol. 49, p. 265, n. 56.

16. Aristotle sees the closest affinity between art and nature in that both proceed by successive stages toward a definite end. In both cases "each step then in the series is for the sake of the next; and generally art partly completes what nature cannot bring to a finish, and partly imitates her." Conclusion: "If, therefore, artificial products are for the sake of an end, so clearly also are natural products." *Physics,* II, 8, 199_a 15-18 [Gilson says "10-18"]. The gradation which reigns in the order of life proves finality for Aristotle, as it proves evolution for Darwin.

17. Bergson, *L'évolution créatrice,* in *Oeuvres,* p. 516 [Mitchell, p. 30]. We have here perhaps an example of the philosophical myth of science, that is to say, of science such as philosophers tend to imagine it. Science itself is more modest and often contents itself with what Claude Bernard called explication, in default of proof. Darwin, exceptionally modest, it is true, was much less sure than Bergson in these matters: "But I believe in natural selection not because I can prove, in any particular case, that it has changed one species into another, but because it groups and explains well (it appears to me) a quan-

tity of facts in classification, embryology, morphology, rudimentary organs, succession and distribution in geology." Fragment of an unedited letter of Darwin's (1861) discovered in the British Museum (Add. Ms. 37725, fo. 6) by Dr. Maurice Vernet and published by him in his book *L'évolution du monde vivant* (Paris: Plon, 1950), preceded by a photographic reproduction of the document. See the same author's *Qu'est-ce que la vie? Quelle est son origine et quelle est sa nature? Conséquences philosophiques que l'on peut en tirer*, an essay published in the work *Humanisme et pensée scientifique* (Paris: Centre économique et social de perfectionnement des cadres, 1969), pp. 18-19. I am very thankful to Dr. M. Vernet for having brought the existence of this document to my attention.

18. René Dubos, "Biological Individuality," *Forum* 12 (1969): 5.

19. Hudson Hoagland, "Biology, Brains and Insight," *Forum* 10 (1967): 27.

20. Dubos, loc. cit., p. 5.

21. Bergson, *L'évolution créatrice*, in *Oeuvres*, p. 532 [Mitchell, p. 51].

22. Ibid., p. 537 [Mitchell, p. 57].

23. The analogy with inspiration in art had already been noted by Paul Janet, *Les causes finales*, 2nd ed., "Preface," p. ix. Analogous views are to be found in G. Seailles, *La génie dans l'art*, cited in *L'évolution créatrice*, p. 518, n. 3 [Mitchell, p. 34, n. 1]. And of course, in Ravaisson, cf. n. 26 below.

24. See Thomas Aquinas, *De potentia*, q. v, a. 1, resp: "*Nam finis non est causa, nisi secundum quod movet efficientem ad agendum; non enim est primum in esse sed in intentione tantum.*"

25. Bergson moreover was conscious of the difficulty of the problem; the biologists did not allow him to ignore it. He cited the remark made in 1897 by the American biologist E. B. Wilson, *The Cell in Development and Inheritance* (New York, 1897), p. 330: "The study of the cell has, on the whole, seemed to widen rather than to narrow the enormous gap that separates even the lowest forms of life from the inorganic world." [Mitchell, p. 44: not cited by Gilson.]

26. Bergson came quite close to a solution, and perhaps even attained it, in his masterful pages of the notice about Ravaisson, *Oeuvres*, pp. 1468-69.

27. Bergson, *L'évolution créatrice*, in *Oeuvres*, pp. 684-85 [Mitchell, p. 244]. The text continues: "Life in its entirety, regarded as a creative evolution, is something analogous; it transcends finality, if we understand by finality the realization of an idea conceived or conceivable in advance." [Mitchell, pp. 244-45.] Yes, but why conceive of it thus?

28. Ibid., p. 698 [Mitchell, p. 261].

29. Ch. du Bos, *Approximations* (Paris: Fayard, 1965), pp. 564-65 and p. 572.

30. "Baudelaire," in Ch. du Bos, *Approximations*, p. 204, n. 3.

31. Baudelaire, *Oeuvres complètes* (Paris, Pléiade), pp. 1029-30.

Chapter V. The Limits of Mechanism

1. Aristotle, *On the Parts of Animals*, I, 1 [642ᵃ].
2. Ibid.
3. William Paley, *Natural Theology*, 1802. We [Gilson] shall cite from the edition of *Natural Theology, or Evidences of the Existence and Attributes of the Deity, Collected from the Appearances of Nature* (London, 1821). This is volume IV of the *Miscellaneous Works of William Paley*, D.D. . . . Rector of Bishop Wearmouth (London, 1821). We are concerned, then, with the work of a theologian, and this quasi-indestructible alliance between the problem of natural teleology, which only concerns the philosophy of nature, and that of the existence of God, which is dependent on natural theology, explains in part the hostility of atheistic biologists toward the notion of the final cause. We shall never hesitate to remember that this alliance is in no way necessary, but it exists, and William Paley is an eminently representative example of it.
4. W. Paley, op. cit., ch. I, p. 14. [See William Paley, *Natural Theology* (London: Hamilton, 1815), pp. 2, 3, 6-8.] Paley next complicates the argument by assuming that in this watch one ascertains in addition the property of making another watch similar to it. This fact orients the argument toward the conclusion that there is a first cause, and with good reason, since it studies finality in the context of a natural theology; but we can imagine a biologist put off by this aspect of the question. Darwin, who had used Paley at the time when he himself was studying for holy orders, retained the impression that the notion of final cause was theological rather than scientific. The convinced finalist, Charles Bonnet of Geneva, was also always a not-less-resolute mechanist: see his *Palingénésie philosophique*, IXᵉ partie, *Réflexions sur l'excellence des machines organiques*, ch. I: "We are not capable of admiring adequately that astonishing display of springs, levers, counterweights, variously calibrated winding, twisting tubing which enters into the composition of organic machines. The inside of the most vile appearing insect engrosses all the conceptions of the most profound anatomist." *Oeuvres complètes*, t. VII, p. 240.
5. For example, Darwin could not think of the problem of the origin of the eye "without feeling a chill"; he had certainly wanted to reply to what Paley said about it (ch. III, pp. 25-26 [pp. 18-19], and the whole chapter). This is what he calls examining the eye "as a piece of mechanism" (p. 36) [p. 35]. Cf. ch. V, 3, pp. 56-59; ch. VI, p. 69. On the mechanical element in the structure of the human body, ch. VIII, pp. 82-107 [pp. 92-122]; the same remarks concerning muscles, blood vessels, etc. . . . Ch. XII on the comparative anatomy of animals is remarkably precise. Darwin will retain, moreover, the wording concerning the woodpecker (ch. XIII, 2, p. 210) [p. 250]. On the comparison of the adaptation of the parts of animals to the adaptation of the parts of a watch, ch. XV, p. 220 [pp. 261-62]. Paley knew Bernardin de Saint-Pierre (XIX, 4, p. 279 [p. 334]; 7, 2, p. 285 [p. 343]) and Erasmus Darwin (XX, p. 298

on *adaptation*) [p. 359], but he foretells Charles Darwin concerning the movement of climbing plants (XX, p. 299) [p. 359]. For his opposition to the theory of the "internal moulds" [*moules internes* – sic] advanced by Buffon, see ch. XXIII, p. 353 [p. 430]; he suspects atheism in this respect (p. 355) [p. 432]. Paley also rejects the doctrine of Lamarck, who wants it to be the case that organs are born from the operations of the organism and that they disappear from lack of use: for centuries the Jews have circumcized themselves but their foreskins have not disappeared (p. 359) [p. 434]. A theological conclusion follows: "Upon the whole, after all the schemes and struggles of a reluctant philosophy, the necessary resort is to a Deity. The marks of design are too strong to be gotten over. Design must have had a designer. That designer must have been a person. That person is God." (p. 363) [p. 441] The work comes to a close with a study of the divine attributes, principally goodness, which requires the discussion of the problem of evil in nature [*du mal naturel*]. It is useful to look at pp. 391-96 [pp. 475-81] concerning the "superfecundity" of species in comparison with what Malthus and Darwin will have to say of it later. [Gilson has just told us (see n. 3 above) that Paley's *Natural Theology* was published in 1802; he has also noted (see p. 137 above; p. 127 in the French text) that Malthus first published his essay *On the Principle of Population* in 1798. Thus his sequencing of events here is partly in error.]

6. "Vitalism has in fact become a 'dirty word' in many circles. This is not what keeps us from it, but rather the fact that if we should try to hold on to it, we would soon be forced to change it beyond all recognition." W.M. Elsasser, op. cit., "Preface," p. v. Before the publication of *Atom and Organism* the author had attempted a first approach to the problem in *The Physical Foundations of Biology* (New York: Pergamon Press, 1958). For a renovated [*renouvelée*] view of the problem posed by vitalism, see the numerous works of Maurice Vernet, notably: *Le problème de la vie* (Plon, 1948); *L'âme et la vie* (Flammarion, 1955); *La vie et son mystère* (Grasset, 1958).

7. Elsasser, *Atom and Organism*, "Preface," v.

8. Ibid., vi.

9. Ibid., vii.

10. Here are the conditions which are necessary in order that a biological theory deserve to be called scientific in the eyes of our author. It ought to admit that: 1.) The fundamental laws of quantum mechanics are applicable to living organisms just as they are to inorganic matter; 2.) life was born progressively on our planet from inanimate matter; 3.) any design for a biological theory which does not allow these conditions, or their consequences, to be satisfied in an entirely natural manner, and not simply by virtue of some artifice, ought to be rejected out of hand [*sur-le-champ*]. We might initially think that these severe restrictions necessarily confirm preconceived mechanist views, but the principal object of the author is to show that this is not the case. Op. cit., p. vi. The same resolution to accept quantum mechanics and the second law of thermodynamics such as they are is reaffirmed on p. 4.

11. Elsasser, op. cit., p. 4.

12. Ibid. [Italics in the original, not in Gilson.] Jacques Monod (*Le hasard et la nécessité* [Paris: Le Seuil, 1970]) passes severe judgment on the thesis of Elsasser and of Polanyi, and less obviously of Nils Bohr himself: "The least one can say is that the arguments of these physicists are oddly lacking in strictness and solidity." (p. 41) [Jacques Monod, *Chance and Necessity* (New York: Knopf, 1971), p. 28.] Nevertheless, neither here nor further along (p. 108) does Monod take Elsasser's principal argument into consideration, namely, the impossibility of a completely mechanical explanation of the heterogeneous. Elsasser does not raise any objection against the possibility of the mechanism of invariance (pp. 41-42) [p. 28]. Jacques Monod himself has to change his tone quite a bit when he comes to the structure of the organism: "This leaves us with teleonomy or, more exactly, with the morphogenetic mechanisms which put teleonomic structures together. It is perfectly true that embryonic development is in appearance one of the most miraculous phenomena in the whole of biology." (p. 42) [p. 28] There follows a denunciation of Elsasser's vitalism which, says Monod, thrives on our ignorance in biology. He next returns to the classic argument of the progress of science which, pushed to the limit, one day will completely eliminate that which remains inexplicable in this domain. Monod does not appear to see that the admirable progress of biological mechanism has left intact the problem of morphogeny, and that even if it should one day attain its term of perfection, mechanist explanation would leave intact the problem formerly posed by Aristotle concerning the origin of the organic. Jacques Monod foresees the day when science will no more leave open to vitalist speculations "the field of subjectivity: that of consciousness itself." (p. 42) [p. 29] But he himself says that "the cornerstone of the scientific method is the postulate that nature is objective. In other words, the *systematic* denial that 'true' knowledge can be got at by interpreting phenomena in terms of final causes, that is to say, of purpose." (p. 32) [p. 21] This postulate is "consubstantial with science"; it is "pure, forever indemonstrable"; and then: "objectivity nevertheless obliges us to recognize the teleonomic character of living organisms, to admit that in their structure and performance (read: *operations*) they act projectively—realize and pursue a purpose. Here, therefore, at least in appearance, lies a profound epistemological contradiction." (p. 33) [pp. 21-22] How can Monod hope to resolve this contradiction, since the postulate of objectivity from which he sets out eliminates one of the two terms from entering into play? In fact, thus understood, the postulate of objectivity is the triumph of subjectivity.

13. [Elsasser, op. cit., p. 11. Gilson does not cite these passages and proceeds quite idiomatically in his paraphrase of them.] *Bonding* in the related sense of *binding*. Without their absolute similarity, which entails their indiscernibility, "no exact meaning could be attributed to a purely statistical distribution of properties of atoms and molecules." Ibid., p. 12. Here is why: *"When the quantum mechanician speaks of a 'system' he always denotes a*

class." (p. 13) [Italics in the English.] A set of atoms or molecules "each having the same composition and all being in the same quantum state, will be denoted as a *fully homogeneous class."* (p. 14)

14. Ibid., p. 14.

15. Ibid., p. 15. Absolutely speaking, we do not know that either; it *appears* evident.

16. Ibid., p. 16.

17. Ibid., p. 20. The author adds: "Similar views were expressed by contemporaries of Bernard who combined physiology with philosophical interests, especially Lotze and Fechner."

18. Ibid., pp. 20-21.

19. Ibid., p. 21.

20. See Georges Canguilhem, *La connaissance de la vie,* 2nd ed. (Paris: Vrin, 1967), pp. 43-80.

21. Ibid., p. 65.

22. Ibid.

23. Ibid., pp. 76-78. See the discovery made by the scrupulous observer Nageotte, of an embryonic tissue preceding by three days the formation of cells in that tissue. Nageotte first of all refused to believe in his own discovery; he rather inferred that the cells had emigrated, but their migration has never been observed. In connection with a controversy provoked in Marxist Russia by Olga Lepechinskaya's book, *Origine des cellules à partir de la matière vivante* (1945), G. Canguilhem exposes a mixture of political interests and scientific convictions. This said, the episode is simply one more in the age-old quarrel between the cellular theory and its adversaries.

24. Elsasser, *Atom and Organism,* p. 33.

25. Loc. cit., pp. 76-77. [Though Elsasser's argument on pp. 76-77 is of this sort, the quotations are on p. 33.]

26. G. Canguilhem, *La connaissance de la vie,* p. 80. Cf. p. 79.

27. Elsasser, *Atom and Organism,* p. 124.

28. Ibid., pp. 123-24. [Italics in the English.]

29. Ibid., p. 124.

30. Ibid., p. 108.

31. Ibid., p. 38.

32. Ibid., p. 44. The physicist naturally looks for a response to this difficulty not in metaphysics but in physics: "We believe this obstacle to lie in the tremendous variability, complexity, and inhomogeneity of organic matter." [p. 44] But that inhomogeneity is what remains to be explained upon the basis of physical mechanism, statistical or not.

33. Ibid., pp. 45-46. Cf. p. 110.

34. Ibid., p. 53. [Italics in the English, not in Gilson.] It is not determinism which is a matter of concern here. Even if one held it absolutely, the problem remains of knowing if it is adequate, under the form of mechanism, to explain the organic [*à rendre raison de l'organique*]. See Etienne Wolf, *Les*

chemins de la vie (Paris: Hermann, 1963), pp. 2-10: "Les critiques du déterminisme et leur valeur," cf. p. 121.

35. Ibid., pp. 58-59.

36. Ibid., p. 134.

37. [Ibid., p. 135.]

38. Ibid., p. 136.

39. [Ibid., p. 137.]

40. Ibid.

41. I leave aside the finalist argument which is foreign to Elsasser's perspective of typewriting apes tapping at random during a quasi-eternity without resulting in reinventing the entire theatrical works of Shakespeare. As with Pascal's wager, it lends itself to long discussions. For example: "This impregnable reasoning has only one fault: it is applicable to any particular event that occurs in the universe, since a priori the probability of such an event is infinitesimal." (J. Monod, "Leçon inaugurale de la chaire de biologie moléculaire" at the Collège de France, November 3, 1967, p, 26.) This is perhaps not to say much. The theatre of Shakespeare is not an *event*; it is an unmeasurable series of ordered and connected events, itself connected to the existence of the English language, the English people, of Shakespeare the individual (for if he had not existed, the theatre that came about without him would not be the theater-of-Shakespeare), and so on in this fashion to eternity. J. Monod is not disquieted by this: "But the universe exists, and it necessarily follows that it should produce events quite equally improbable, and man find himself to be one of them. He drew the first prize." (p. 26) All events are equally improbable only if they are all of the same nature, which is questionable. Besides, weak as it may be, the probability of winning first prize is never nothing, for it is a probability, whereas in the lottery in question the first prize does not exist. With respect to DNA [ADN], that "philosopher's stone of biology" (p. 12), since it is "by itself inert and devoid of teleonomic qualities" (p. 16), it does not explain any teleology; it is rather that teleology has the task of explaining it. Finally, to have recourse to notions of information and "molecular communication" (p. 21) in order to give an explanation of the morphogenetic fecundity of mechanism is to use linguistic metaphors without explaining anything.

Chapter VI. The Constants of Biophilosophy

1. Paul Lemoine, in the *Encyclopédie française*, t. V, 08-2.

2. Paul Lemoine, "Du vitalisme au finalisme," *Encyclopédie française*, t. V, 08-2. Speaking of finalism and vitalism as if they were two varieties of the same doctrine, this scientist says: "These tentative explanations were driven out of physiology in the first half of the nineteenth century by those who put that science on the ground where it develops today. Notable among them was

Claude Bernard. But morphology, and above all biology in the broad sense of the word, have remained as 'redoubts' for finalism where it still maintains itself under a more or less rejuvenated aspect. Nor is a suspicion of latent, and, so to speak, occult finalism absent from the protagonists of evolutionary theories themselves." Claude Bernard certainly drove finalism out of physiology but, we have seen, not out of biology "in the broad sense of the word." With respect to Darwin, we have also seen that, on the contrary, he was flattered by the thought of having reconciled finalism and mechanism. Likewise with Lamarck, if, as Paul Lemoine himself observed, adaptation to the environment, like natural selection, implies a sort of finality.

3. Georges Canguilhem, *Etudes d'histoire et de philosophie des sciences* (Paris: Vrin, 1968), pp. 306 and 323.

4. Claude Bernard, *Cahier de notes (1850-1860)*, ed. M.D. Grmek (Paris: Gallimard, 1965), pp. 58-59. Bernard, for all that, does not consider research into final causality as nonscientific. (p. 84) That is not in his eyes a reason for denying its existence in nature. Cf. "Individuality rules teleology. The formation of an individual, a complete organism which has its entelechy, etc., is the final evolutionary goal." Op. cit., p. 200.

5. "Life is an evolution." Ibid., p. 154. Cf. pp. 230-31. The life in question is that of the individual, ontogenesis.

6. Paul Lemoine, *Encyclopédie française,* t. V, 82-11.

7. For example: Man will create at his pleasure living beings, "not only those [species] which exist or which have existed, but others still which will be endowed with qualities which man shall desire. For life is one of the rare forms of energy–the most supple, perhaps–which Man has not yet known how to bend to his will." Paul Lemoine, *Encyclopédie française,* t. V, 82-11. The first two species created will probably be slaves and soldiers. In any case we would not know how to predict happy prospects for the future creatures of man. Cf. the appropriate [*juste*] remarks of Etienne Wolf, *Les chemins de la vie,* pp. 177-95, and Jean Rostand, op. cit., "Preface," pp. xiv-xv.

8. Pierre-P. Grasse, *Encyclopédie française,* t. IV, p. 1. The intransigent dogmatism of the mechanists, moreover, leads one erroneously to believe that all biologists are antifinalists. Proof to the contrary will be found in Lucien Cuénot, *Invention et finalité en biologie* (Paris: Flammarion, 1941). Beyond his own conclusions see the witnesses he has assembled on pp. 45-47: Lippman, Guye, Lecomte de Nouy, Léo Errera, Gagnebin, Conklin, R. Broom, Ch. Richet. Lucien Cuénot himself insists on the inventive activity at work in nature. Whatever may be the cause of it, it exists and produces genuine tools. The pincers of a lobster do not resemble genuine pincers; they are genuine pincers. "Natural teleology is not a theoretical interpretation; it is the most incontestable of facts." Op. cit., p. 40. Let us note, moreover, that to call it [natural teleology] "antichance" (pp. 48-49) is to define it by opposition to something which is nothing. Chance is a by-product of order, not the other way around.

9. We can see in Lucien Cuénot's book, all of whose sympathies are with finalism, that no scientifically satisfying explanation of it appears to have been found: op. cit., "The nonmechanist or finalist theories," pp. 121-53. He rightly concludes that from these abortive attempts emerges nevertheless "a common metaphysic" (p. 152). In all the antimechanist attempts he recounts "it appeared necessary to lodge in the Cartesian machine an inventor-conductor; the Lamarckists, mnemonists, entelechists, holists, organicists, try to express an irrational, doubtless, inexpressible, metaphysical entity which they imagine in it: vital principle, autonomy of life organoformative idea, organic intelligence, psychoid, cellular conscience, totality of concept, entelechy, *élan vital*, etc. . . . Basically, these obscure words are symbols of the profound unknown cause of which we have need in order to interpret biological finality." Op. cit., p. 153; cf. p. 44.

I owe it to the excellent book of P.-H. Simon (*Questions aux savants*, ch. III) that I know of a text of Louis de Broglie which appears to me to summarize perfectly the really scientific position of the present question: "It seems unbelievable that similar organs (the eye, the ear of higher animals, etc.) can have been produced by the sole agency of chance, even though it be prolonged over enormous periods of time. Life's realizations seem to come about from an organizing force which does not manifest itself in inert nature and whose true nature appears entirely unknown to us." *Revista Euclides*, vol. XI, May–June 1957; in P.-H. Simon, op. cit., p. 98, n. 1.

10. Julian Huxley [*Evolution in Action* (New York: Harper, 1957), pp. 11-12], cited by John C. Greene, *Darwin and the Modern World [View]* (New York: Mentor, 1963), p. 71. Notice the American tendency to take evolutionary Darwinism for a phenomenon marking off an age and of planetary significance.

11. Emile Guyenot, cited by Marcel Prenant (*Biologie et marxisme*, E.S.I. [Paris, 1936]) and by Lucien Brunelle in his introduction to *Lamarck, pages choisies* (Paris: Editions sociales – Les classiques du peuple, 1957), p. 38. Not having felt the obligation of entangling myself in the maze of Marxist evolutionism since I cannot see how the economic order has been able to influence scientific thought on this point, I shall rest content with sending the reader to this little book where the contrary point of view is set forth with perfect straightforwardness. The author thinks that, without wishing to force his conclusions upon science, the rational explanation entailed "seems to him to have to take place rather by means of the thoroughness of adaptation than by means of natural selection." (p. 40) It only remains then [*il ne restera plus alors que*] to explain how the official scientific truth of Marxism can be traced back to Jean-Baptiste-Pierre-Antoine de Monet, Marquis de La Marck, son of the Lord of Bazentin in Picardy, of a family whose nobility can be traced back at least to the time of Henry IV. It is true that he was poor, but, to the point, Chateaubriand taught us that that which is most precious in nobility is to maintain the principle that there is something more important than money.

Chance as a positive constructing agency has recently found a passionate partisan in Jacques Monod, *Le hasard et la nécessité* (Paris: Le Seuil, 1970) [*Chance and Necessity* (New York: Knopf, 1971)]. This biochemist (he is not a zoologist) puts his confidence in Darwinian natural selection brought to completion by the discovery of DNA. "Drawn out of the realm of pure chance, the accident enters into that of necessity, of the most implacable certainties. For natural selection operates at the macroscopic level, the level of organisms." (p. 135) [p. 118] The problem of the origin of species becomes henceforth "the major problem" of "the origin of the genetic code and of its translation mechanism. Indeed, instead of a problem it ought rather to be called a riddle." (p. 159) [p. 143] [Nothing is said in the English translation about the origin of species. The "it" in the quotation above refers to the origin of the genetic code, not the origin of species. The French original reads: *"Mais le problème majeur, c'est l'origine du code génétique et du mécanisme de sa traduction."* (Monod, p. 159) Gilson says: *"Le problème de l'origine des espèces devient désormais 'le problème majeur de l'origine du code génétique'."*] Chance is the major element of the response. The demonstration of this point is only a long paralogism wherein chance as the *condition* of all possible teleonomy is confounded with chance as *cause* of all teleonomy: "Chance alone is the source of every innovation, of all creation in the biosphere. Pure chance, absolutely free but blind, at the very root of the stupendous edifice of evolution: this central concept of modern biology . . . is today the *sole* conceivable hypothesis, the only one that squares with observed and tested fact." (p. 127) [pp. 112-13] The classic examples (p. 128) [p. 114] of a passerby who is killed by the chance slippage of a tile from a roof [it is a workman's hammer in the English translation] explains the death of the passerby well, but one would love to see tiles reorganize themselves so as to stop up the holes in a roof. We ask ourselves how this scientist can reconcile the "teleonomic performances of the proteins," founded on their *stereospecific* properties (p. 60) [p. 46], with the thesis of pure chance as the origin of organisms. The remarkable structure of DNA itself requires an origin. The word "chance" being nothing more than the sign of a causal deficiency [*carence causale*], it is astonishing that a scientist could take as physical causes the accidental variations which are only the points of possible insertion for such causes. La Bruyère's adage comes to mind: "The mind is worn down like everything else; the sciences are its sustenance, they nourish it and consume it."

12. G.G. Simpson, "Biological Sciences," in *The Great Ideas Today* (1965), pp. 300, 311, 317 [sic].

13. Theodosius Dobzhansky, *Heredity and the Nature of Man* (New York and Toronto: Signet, 1966, p. 118 [New York: Harcourt, Brace and World, 1964, p. 115]: "Nothing makes sense in biology except in the light of evolution," understood according to the Darwinian spirit; he describes it nevertheless while saying: "the diversity of living beings is a response of living matter to the diversity of the environments on earth" [1964; pp. 115-16], which

is to define evolution as Lamarck did, and to overlook the remark of Claude Bernard that it is not a question of living matter but of living beings. Finally, the same biologist says that evolution is "chance," which leads us to the simple *panta rei* of Heraclitus, with, however, this addition, that organisms have always evolved from more simple to more complex forms (loc. cit.). This colossal generalization, contested by many a zoologist, is due to Spencer, not to Darwin. Nothing shows better the indifference to precision of which certain scientists give proof when they step outside the limits of their science. But there is a natural bit of chicanery in this indifference to truth. What scientist would concern himself with calling upon Spencer as a witness? One then spontaneously attributes that notion of the philosopher Spencer to the scientist Darwin in the hope of transforming it into a scientific idea. But the idea remains what it is.

14. Sainte Beuve, *Oeuvres* (La Pléiade), t. II, p. 117.

15. Organic teleology is a teleology of *fact,* "not only morphologically but also functionally: a crab's pincer is indeed an effective pincer," Lucien Cuénot, *La finalité en biologie* (Paris: Hermann, 1948), in the collection, Actualités scientifiques et industrielles, no. 1067, p. 40. On the same page we find an impressive list of devices included in the structure of animals: saws, knives, pressure-buttons, etc. It is for a properly scientific reason that Cuénot did not believe in Darwinian natural selection: "no known mutation could be considered as the beginning of a tool or the development [*coaptation*] of an organ toward a useful function." (p. 42) And then, "selection [*triage*] does not exist; death kills randomly." (p. 45) Among the millions of Jews massacred under Adolph Hitler's instructions, how can we know whether there was another Spinoza, another Einstein? At the purely material level, what choice can be exercised over the future living beings which abortion kills before birth, or whose conception science, supporting [*secondant*] nature today, prevents?

16. George G. Simpson, "Biological Sciences," loc. cit., p 309, n. 9. Selection can be perhaps an antichance factor, but it is understood as if it did not "choose," as operating on the small number of survivors of a massacre of immense extent and in which death is not a matter of choice, and, finally, that it only works [*s'exerce*] on individual variations, which are in the realm of chance. See, moreover, the remark of Julian Huxley, n. 10 above.

17. L. Cuénot, *La finalité en biologie*, p. 38. In "Face à Face: Pierre-Henri Simon et Jacques Monod" (*Atomes*, no. 268, September 1969, p. 481), the latter sets forth three senses of the word "chance," of which the second is exactly that which Aristotle gives to it. Remembering Darwin, he adds that "chance is in the structure of the DNA, necessity is in selection" (p. 481). We are going around in circles (*nous sommes au rouet*). First of all, as P.-H. Simon pertinently asks, is the DNA itself a product of chance? Next, if we admit that the first spontaneous variation, which Darwin and Wallace place at the origins of the transformation of the species, is indeed the result of chance, it is because we do not know a final cause for it. "Selection is not a phenomenon

of chance. . . . From this moment we are in [the realm of] macroscopic necessity." (p. 481) This is to avoid saying how an order of parts not at all macroscopic can issue from necessity alone. Everything is necessary in music except the music. In fact, Jacques Monod does not see any other factor than necessity in the explanation of the evolution of species (p. 482), but not to see any other explanation does not prove that what one does see is scientific, above all if it is not an explanation. To say that chance is alleged not as an explanation but as a "fact," and to add that "this is completely different" (p. 483), is simply to repeat the same thing, for chance is nothing but the absence of explanation. The absence of reality and intelligibility is not a fact; it is nothing.

On M. Jacques Monod's positions one would read with profit what appears to me to be the quite faithful exposition and, to my sense, the pertinent critical remarks of P.-H. Simon, *Questions aux savants* (Paris: Edition du Seuil, 1969), particularly chapter III, "On Life as a Phenomenon and as a Prodigy."

18. We reread with interest the article "Fin, Causes finales" in Voltaire's *Philosophical Dictionary*. Resolutely *"causefinalier,"* Voltaire has very well discerned at least two of the principles which distinguish true teleology from false, namely, that the effects be independent of the beings to which they belong, and that they be always the same, at all times and everywhere. Op. cit., ed. Garnier-Flammarion (Paris, 1964), p. 192. [See Voltaire, *Philosophical Dictionary* (Penguin Books, 1971), p. 205.]

19. Lucien Cuénot, *La finalité en biologie*, p. 38.

20. The refutation of the theological finalism proper to Leibniz ought not be taken as a refutation of natural finalism. Inversely, if the presence of a quite real muddle [*gâchis*] in nature (L. Cuénot, *L'évolution biologique*, p. 367) is a valid argument against the notion of a creator God (except of a limited God [*Dieu fini*] who does what he can, and even bungles things), the five to seven million eggs produced annually by the cod do not prove that there may not be natural teleology. Even poorly constructed, a lock presupposes a locksmith. The recognition of final causes does not imply that final causality works perfectly, but that it may exist. Where there is no order, there can be no muddle.

21. De Maupertuis, "Accord de différentes lois de la nature qui avaient jusqu'ici paru incompatibles," *Mémoires de l'Académie royale des Sciences*, 1744, p. 425. The memoir is on pages 417-26.

22. We can only subscribe to the wise conclusion of the biologist Lucien Cuénot in *L'adaptation* (1925), pp. 395-96: "The biologist, whether it be in his spiritual conscience (for Cuénot one is either a spiritualist or a materialist) or his atheistic conscience, then has to consider final causality only as a fact. He only has to study its determinism, its strivings, its errors, if there are any, exactly as a physicist or a chemist who studies the phenomena of his speciality. If the known factors of evolution and adaptation appear to him insufficient, he has but to recognize his ignorance and appeal to a future better informed about the number and value of efficient causes. For him final causality

is immanent, that is to say that the being in which one notes connections of means to ends is also the activity which realizes that end by these means" [*sic*]. It is impossible to make the exact nature of the Aristotelean *telos* more intelligible. L. Cuénot gave, on pp. 397-407, a good bibliography on the subject up to 1925. The philosophy of nature calls upon final causality in order to explain the structure of living beings. The metaphysician invokes the notion of God to explain the existence of final causality. These are two different and even distinct problems. The first one only allows the second one to be posed; it is not competent to resolve it.

23. François Jacob's remarkable book, *La logique du vivant: Une histoire de l'hérédité* (Paris: Gallimard, 1970) could be taken to mark the climax of the modern history of finalism. Its author takes it to be a weapon against religion (see the initial inscription) ["Do you see this egg? With it you can overthrow all the schools of theology, all the churches of the earth." Diderot, *Conversation with d'Alembert.*] He takes exception to vitalism and animism ("no psyche to direct operations, no will to prescribe their pursuit or stoppage"; "an internal logic which no intelligence has chosen," p. 318), after which, contrary to all expectations, this biochemist declares: "we see how this attitude differs from the reductionism which has long prevailed," p. 321. We are not astonished then when he uses finalist language: end (pp. 291, 297), project (p. 307); he does not hold to chance as a sufficient explanation of the origin of life (p. 326); with courage, he writes: "To recognize the teleology of living systems is to say that one can no longer work at biology without referring constantly to the *project* of organisms, to the sense that gives existence even to their structures and functions." And further: "Up to the present . . . the rigor imposed on description necessitated the elimination of that element of teleology which the biologist refused to admit in his analysis. Today, on the contrary, it is no longer possible to dissociate the structure from its meaning, not only in the organism but in the series of events which have led the organism to be what it is." Finally: "natural selection imposes a finality not only on the organism in its entirety but on each of its constituent parts," etc., pp. 321-22. [I have not been able to locate all the items M. Gilson cites. For the last passage see François Jacob, *The Logic of Life*, tr. Betty Spillmann (New York: Vintage, 1976), p. 300. The translations here are my own except for the initial inscription.] It is interesting to read the precise details concerning related problems, pp. 323-27. We will not be too rigorous in our attitude toward this biologist who is anxious to avoid all metaphysics by transposing his biology into terms of linguistics, which is his privilege, but which makes of science a perpetual metaphor. The cell is not a *text*, because there is no one to write it; it is not composed of *signs*, because its elements are not letters destined to be read; it does not constitute a *message*, because it does not contain any intelligible sense to be communicated by one mind to another which might be there to receive it. This is, Aristotle said, "to make poetical metaphors." Provided that we do not take them for science, nor even for philosophy, it would be wrong to deny ourselves the use of them.

Appendix II: Darwin in Search of Species

1. [Charles Darwin, *The Origin of Species*, ch. VII, p. 173 (112). The edition of the *Origin* used is apparently the sixth, the last published during Darwin's lifetime (see p. 2 for indirect evidence of this), though this is not specified. Gilson uses the edition of *The Origin of Species* and *The Descent of Man* to be found in volume 49 of The Great Books of the Western World, ed. R.M. Hutchins and M.J. Adler (Chicago: Encyclopaedia Britannica, 1954): see note 1 in the second section of Gilson's chapter 3 (p. 82 of the French; p. 100 above). As mentioned there in our notes, the pages have been correlated, the Great Books edition pages given first after the entry *"Origin,"* with the Modern Library pagination following in parentheses. All pagination has been provided by the translator.]

2. Like Darwin, Lamarck earlier wavered between the reality and the unreality [*irréalité*] of species. They are creations of the mind:

"But these groupings . . . are altogether artificial, as also are the divisions and subdivisions which they present. Let me repeat that nothing of the kind is to be found in nature, notwithstanding the justification which they appear to derive from certain apparently isolated portions of the natural series with which we are acquainted. We may, therefore, rest assured that among her productions nature has not really formed either classes, orders, families, genera or constant species, but only individuals who succeed one another and resemble those from which they spring. Now these individuals belong to infinitely diversified races, which blend together every variety of form and degree of organization; and this is maintained by each without variation, so long as no cause of change acts upon them." Lamarck, *Philosophie zoologique*, ed. Ch. Martins (Paris, 1873), t. I, ch. I, p. 41 [Elliot, pp. 20-21]. But a bit further on, Lamarck asks that we study "the natural method," that is, that our researches "conform to the exact order found in nature, for that order is the only one which remains stable, independent of arbitrary opinion, and worthy of the attention of the naturalist." Op. cit., t. I, ch. I, p. 43 [Elliot, p. 22]. The foundation of his thought is that species is a provisionally stable state between two mutations whose stability is tied to the stability of its conditions of existence: "species . . . have only a relative constancy, and are only invariable temporarily." (p. 90) [Elliot, p. 44] It is useful to name species as "any collection of like individuals, perpetuated by reproduction without change, so long as their environment does not alter enough to cause variations in their habits, character, and shape." Op. cit., p. 91 [Elliot, p. 44].

3. [*Origin,* p. 29 (p. 46).]

4. [*Origin,* p. 27 (p. 43).]

5. [*Origin,* p. 27 (p. 43).]

6. [Although the thought of Buffon concerning species can be accused of being inconsistent, Gilson's comment here is apparently inconsistent with his earlier remark distinguishing Buffon and Lamarck from all the other predecessors of Darwin on the precise issue of the stability of species.]

7. [*Origin,* p. 28 (p. 45).]

8. [*Origin,* p. 53 (p. 83).]

9. [*Origin,* p. 159 (p. 246). This is chapter 10 in the sixth edition, the one reprinted in the Great Books edition Gilson claims to be using (see n. 1, p. 98, above) and in the Modern Library edition. It was chapter 9 through the fifth edition. Gilson may be using Peckham's *Variorum* text here.]

10. [*Origin,* (ch. II), p. 26 (p. 41).]

11. [Gilson erroneously calls it ch. 8. Gilson is apparently using Peckham's *Variorum* text, or the fifth edition of the *Origin,* here also. The chapter on "Hybridism" is chapter 8 of the *Origin* until the sixth edition, when it becomes chapter 9. At any rate, Gilson is not using the Great Books edition here.]

12. [*Origin,* p. 136 (pp. 209-10).]

13. [*Origin,* p. 137 (pp. 210-11).]

14. [*Origin,* p. 139 (p. 214).]

15. [*Origin,* pp. 242-43 (p. 372).]

16. [*Origin,* p. 243 (p. 373).]

17. An American philosopher has praised "the great theologian Sertillanges, with having (in 1945 and at other times) in vain, protested against the artificial opposition maintained between the notions of evolution and creation." In fact, "nothing prevents us from seeing in evolution, instead of a substitute for creation, simply another perspective on the manner in which the creative fact [*creative fact* for *act?*] [sic] . . . is linked to the facts of nature." Lamarck never intended his own theory to be understood otherwise. "Any collection of like individuals which were produced by others similar to themselves is called a species. This definition is exact. . . . But to this definition is added the allegation that the individuals composing a species never vary in their specific characters, and consequently that species have an absolute constancy in nature. It is just this allegation that I propose to attack, since clear proofs drawn from observation show that it is ill-founded." Lamarck, *Philosophie zoologique*, ed. Ch. Martins (Paris, 1873), t. I, ch. III, p. 72 [Elliot, p. 35]. "It has been imagined that every species is invariable and as old as nature, and that it was specially created by the Supreme Author of all existing things." Op. cit., I, III; t. I, p. 74 [Elliot, p. 36]. "I shall then respect the decrees of that infinite wisdom and confine myself to the sphere of a pure observer of nature. If I succeed in unravelling anything in her methods, I shall say without fear of error that it has pleased the Author of nature to endow her with that faculty and power." Loc. cit., t. I, pp. 74-75 [Elliot, p. 36]. As they say at times in America: "That is God's way of doing things." We recall only that the conclusion of the *Origin of Species* entirely agrees with this view.

18. [*Origin,* p. 243 (p. 374).]

19. [Gilson erroneously specifies the thirteenth.]

20. [*Origin,* p. 208 (p. 319).]

21. [*Origin,* p. 208 (p. 319).]

22. [*Origin,* p. 18 (p. 30).]

23. [Gilson erroneously says ch. 6.]

24. [*Origin,* p. 44 (p. 69). The first sentence Gilson quotes is not in the sixth edition (1872) of the *Origin.* The *Variorum* says that this sentence was eliminated in the fifth edition.]

25. [This should probably be "chapter I."]

26. ["Unconscious selection"–used in ch. I; see p. 19 (p. 32).] In many a phrase Darwin astutely adds a word to suggest that the stockbreeders had done that "unconsciously." For example: "King Charles's [spaniel] has been unconsciously modified [to a large extent] since the time of that monarch," p. 109 [not to be found there; p. 20, ch. 1, 6th ed. (p. 32)]; apropos the English pointer: "what concerns us is, that the change has been effected unconsciously and gradually," p. 110 [not there; p. 20 (p. 32), 6th ed.]; "in this case there would be a kind of unconscious selection going on," p. 111 [not there; p. 20 (p. 33), 6th ed.]; the melting pear, so different from the wild one: "the art . . . has been followed almost unconsciously," p. 111 [not there; p. 21 (p. 33), 6th ed.]; this is *almost* priceless, for, all the same, Darwin is not sure that the melting pear has been obtained by a series of choices *completely* unconscious! At one time Darwin emphasizes that stockbreeders do not have a precise image of the goal they strive for; at another time, on the contrary, "the animal or plant should be so highly valued by man, that the closest attention is paid to even the slightest deviations in its qualities or structures," p. 115 [not there; p. 22 (p. 36), though Gilson's French does not quite match this passage]. Summing up, an extreme attention to the least variation without attending to the end pursued [is necessary]. The scientific rigor of this process of reasoning is not extreme, but its nonchalance is quite Darwinian.

27. [*Origin,* (ch. I), pp. 18-19 (p. 30).]

28. [Ibid.]

29. [*Origin,* p. 19 (p. 30).]

30. [*Origin,* p. 41 (p. 65).]

31. [*Origin,* p. 41 (p. 65).]

32. [*Origin,* p. 41 (p. 66).]

33. [*Origin,* p. 41 (p. 66).]

34. The analogy between domestication and natural selection struck Lamarck before Darwin: "Now if a single case is sufficient to prove that an animal which has long been in domestication differs from the wild species whence it sprang, and if in any such domesticated species, great differences of conformation are found between the individuals exposed to such a habit and those which are forced into different habits, it will there be certain that the first conclusion [that nature or her author foresaw all the circumstances in which a species would live] is not consistent with the laws of nature, while the second [Lamarck's–i.e., that nature produced serially all species from the simple to the complex, developing habits and organs as the enviroment dictated], on the contrary, is entirely in accordance with them." *Philosophie zoologique*, 1re partie, ch. VII [Elliot, p. 127], cited with the approbation of

Lucien Brunelle, *Lamarck*, p. 96, note 2, as prefiguring the work of Daniel in France, and, in Russia, of Mitchourine, who at the end of his career had the opportunity "of receiving all the assistance he could wish for from the Soviet government for whom the union of theory and practice constitutes a golden rule." [The passage cited by Gilson does not come directly from Lamarck but must be found in Brunelle. I have given the text as translated by Elliot.] On the contrary, for reasons connected with his conception of the history of science, M. Camille Limoges (*La sélection naturelle*, pp. 101, 147-48) underlines the fact that natural selection was conceivable, and had been conceived by Darwin, without the assistance of this accessory pedagogical "model." It is certain that the analogy with domestication is a necessary element in the doctrine such as Darwin himself conceived it. Domestication is the only empirical evidence [*fait*] given upon which the theory could be founded: hence the important work of Darwin's published nine years after the *Origin*, the *Variations of Animals and Plants under Domestication*, 2 vols. (London: Murray, 1868). As Darwin himself put it precisely in the "Introduction" to the *Origin*: "I shall devote the first chapter of this Abstract to Variation under Domestication. We shall thus see that a large amount of hereditary modification is at least possible." [*Origin*, p. 7 (p. 12).] And he has no other proof of the matter, for natural selection is a theory, whereas under domestication one sees hereditary modification still at work today. It was for him one of those "agencies which we see still at work."

35. [*Origin*, pp. 43-44 (p. 69). Gilson has taken liberties with the text in his "quotation." His French reads: "Parlant du 'brutal' amateur de combats de coqs, qui n'est sans doute pas un savant biologiste, Darwin observe 'qu'il sait bien qu'il peut améliorer sa race en choisissant avec soin les meilleurs coqs', ce que l'on tiendra difficilement pour de la sélection inconsciente." I have tried to reconstruct this passage as intelligibly and as simply as possible, while remaining true to Darwin's text.]

36. [*Origin* (6th ed.), p. 44 (p. 70).]

37. [*Origin*, p. 45 (p. 71). Emphases are Gilson's.]

38. [*Origin*, pp. 49-50 (p. 78). Emphases are Gilson's. Gilson paraphrases this text.]

Index of Names

Subject Index